PENGUIN CANADA

THE DAILY PLANET BOOK OF COOL IDEAS

JAY INGRAM has been the host of Discovery Channel Canada's *Daily Planet* since it began in 1995. At the time it was the only hour-long, prime-time daily science show in the world. Prior to joining Discovery, Jay hosted CBC radio's national science show, *Quirks and Quarks,* from 1979 to 1992. During that time he won two ACTRA awards, one for best host, and several Canadian Science Writers' awards. He wrote and hosted two CBC radio documentary series and short radio and television science stories for a variety of programs. He was a contributing editor to *Owl* magazine for ten years, and wrote a weekly science column in the *Toronto Star* for twelve. Jay has also written ten books.

Jay has received the Sandford Fleming medal from the Royal Canadian Institute for his efforts to popularize science, the Royal Society's McNeil Medal for the Public Awareness of Science, and the Michael Smith award from the Natural Sciences and Engineering Research Council. He is a Distinguished Alumnus of the University of Alberta and has received four honorary doctorates.

THE DAILY PLANET BOOK OF COOL IDEAS

Global Warming and What People Are Doing About It

JAY INGRAM

PENGUIN CANADA

PENGUIN CANADA

Published by the Penguin Group

Penguin Group (Canada), 90 Eglinton Avenue East, Suite 700, Toronto, Ontario, Canada
M4P 2Y3 (a division of Pearson Canada Inc.)

Penguin Group (USA) Inc., 375 Hudson Street, New York, New York 10014, U.S.A.
Penguin Books Ltd, 80 Strand, London WC2R 0RL, England
Penguin Ireland, 25 St Stephen's Green, Dublin 2, Ireland (a division of Penguin Books Ltd)
Penguin Group (Australia), 250 Camberwell Road, Camberwell, Victoria 3124, Australia
(a division of Pearson Australia Group Pty Ltd)
Penguin Books India Pvt Ltd, 11 Community Centre, Panchsheel Park, New Delhi – 110 017, India
Penguin Group (NZ), 67 Apollo Drive, Rosedale, North Shore 0745, Auckland, New Zealand
(a division of Pearson New Zealand Ltd)

Penguin Books (South Africa) (Pty) Ltd, 24 Sturdee Avenue, Rosebank, Johannesburg 2196, South Africa

Penguin Books Ltd, Registered Offices: 80 Strand, London WC2R 0RL, England

First published 2008

2 3 4 5 6 7 8 9 10 (CR)

LIBRARY AND ARCHIVES CANADA CATALOGUING IN PUBLICATION

Ingram, Jay
The Daily Planet book of cool ideas : global warming and what
people are doing about it / Jay Ingram.

Includes index.
ISBN 978-0-14-316935-2

1. Global warming.   2. Environmental protection—Citizen participation.
I. Title.

QC981.8.G56I48 2008   363.738'74   C2008-902636-5

Visit the Penguin Group (Canada) website at **www.penguin.ca**

Special and corporate bulk purchase rates available; please see **www.penguin.ca/corporatesales**
or call 1-800-810-3104, ext. 477 or 474

THE
DAILY PLANET
BOOK OF
COOL IDEAS

Contents

Introduction

IT IS UNDENIABLE THAT THE WORLD has warmed and cooled over its history. It's also undeniable that it is warming now. The question is, is it now warming more rapidly than ever before? If so, you might suspect that this is something more than a natural process, and that we have something to do with it. Can we attribute climate change to our emissions of carbon dioxide? If so, then the onus is on us to do something about it. So we need to start with the data that exist, and the conclusions that can be drawn from them.

Fifteen thousand years ago, most of Canada was covered by unimaginably thick sheets of ice. Only rare places, like the eastern side of the Rocky Mountains and a narrow strip down the west coast, were ice free. In fact, those corridors were crucial for the migration of wildlife— including humans—to North America from Russia, via the Bering land bridge. The only reason this pedestrian walkway over what we now know as the Bering Strait existed was that so much sea water was locked into glaciers worldwide that sea levels dropped and exposed the ocean floor.

▶ Ice is the symbol of global warming. The world's glaciers are receding, their ice melting and flowing away. In the Antarctic, that's enough ice to raise sea levels— by how much, no one knows. Thousands of feet above sea level, the glaciers of the Himalayas (seen here) are retreating up the mountainsides, their pace accelerated by sunlight-absorbing pollutants from industrial Asia.

It is the *speed* of warming that has convinced the vast majority of climate scientists that something is going on, and that something is *us*.

We are actually still in the ice age of which that last major glaciation was a part. (Ice ages are not homogeneous: during any one of them glaciers advance, then retreat as the earth warms and cools.) The last glaciation lasted almost one hundred thousand years. At its height, about eighteen thousand years ago, all of Canada and the northern United States was submerged under ice, some of it 3 kilometres thick. Ice ages—and the warmings and coolings within them—go back billions of years.

Of course, there have also been long periods of much warmer temperatures than we now have. Climate change is the norm. This is inarguable, and it is a fact that the small number of scientists who don't believe that we are altering the climate emphasize. They have a point: if climate has changed in the past, why are we so concerned that it is changing now?

Because it's not just that it's changing—it's changing so quickly. It is the *speed* of warming that has convinced the vast majority of climate scientists that something unusual is going on, and that something is *us*. Fair enough, but how do you decide when warming is happening faster than it should?

You have to know what climate has done in the past—how it behaved not just a few generations ago, but a thousand years ago, or even a hundred thousand. The problem is that up until the 1800s, nobody systematically recorded climate data. So scientists have been forced to turn elsewhere to figure out what has gone on. But it's not an easy record to put together.

TAKING ENGLAND'S TEMPERATURE

The oldest ongoing record of temperatures is the Central England Temperature Record, which goes back—incredibly enough—to 1659. Covering a triangular patch of southern England, the record was at first monthly, but became daily in 1772. It is the longest instrumental temperature record in the world. That's impressive, but obviously the flip side is that there are no adequate, accurate temperature records globally for anywhere near that span of time. It was only in the 1850s that decent records began to be kept in a variety of locations around the world. A hundred and fifty years of temperature records are useful but nowhere near sufficient for reconstructing the history of climate.

ART GALLERIES

In the absence of actual data, we can infer what the climate was like before instrumental record-keeping from written descriptions of the weather, like unusual snowfalls, floods, or droughts. These sources can be indirect: for instance, the summer temperatures in Paris from the 1300s to the 1800s have been gauged from written records of the grape-harvesting times. Artworks, like the painting on page xii, depicting skaters on the Thames River in what is called

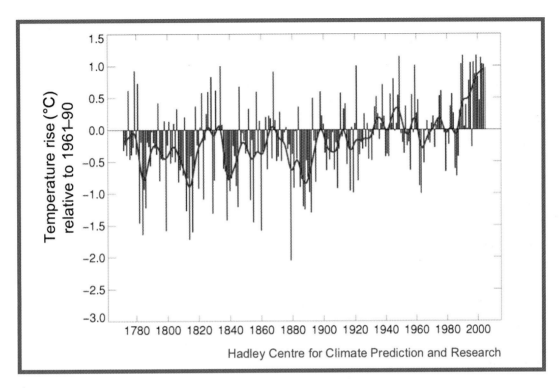

Temperature rise (°C) relative to 1961–90

Hadley Centre for Climate Prediction and Research

▲ No matter how records of past temperatures are assembled, it's clear that there has been unusual warming over the last few decades.

the Little Ice Age, also give us clues to the weather of the past. But these works don't provide anything like a detailed picture, and we can't even be sure they're accurate.

TREE RINGS

Tree rings are good records—the wider the ring, the warmer and wetter the year. Younger trees can overlap with trees that died centuries ago, and so the tree ring record can be pushed back further and further, to beyond the beginnings of recorded history in some parts of the world.

▲ Tree rings have two advantages as a record of climate: they're relatively easy to read, and records from recent trees can overlap with older ones, pushing the record further back in time.

▼ *A Frost Fair on the Thames at Temple Stairs*, painted by Dutch artist Abraham Hondius, probably in 1684. The winter of 1683–84 was incredibly severe, but Londoners, handed a climatic lemon, made lemonade. There were booths, rides on a wheeled boat, coaches, and horses, all testaments to the thickness of the ice.

ICE CORES

Ice cores are tubes of ice drilled out of glaciers. The accumulation of snow on top of the glacier is seasonal, so these cores have their own version of tree rings. The tiny bubbles of gas, locked in when the ice froze, reveal both the composition of atmosphere and, by the ratio of their isotopes, the temperature *when that ice formed*. They are a precious resource.

Even the tiny shells of long-dead ocean creatures and the kinds of pollen grains found in ancient deposits can reveal what the temperature was when they formed.

It's very tricky to assemble this patchwork of indicators into a complete picture of past climate, but when that's done, it's clear that global temperatures rose unusually fast in the last half of the twentieth century. The records of the last decade or so not only include several of the warmest years since records have been kept, but are likely to be the warmest of the last one thousand years—maybe even the last two thousand.

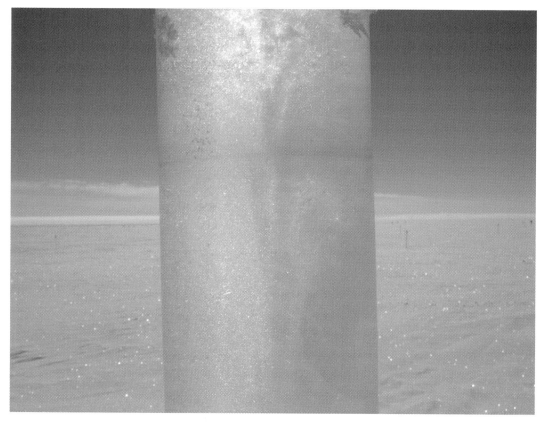

▲ Ice cores are trickier than tree rings but contain a wealth of ancient climate information.

THE HOCKEY STICK GRAPH

The iconic piece of Canadian sports equipment was dragged into the climate debate when scientists published a graph in 1998 that they claimed tracked global temperatures over the last thousand years. The graph had an upward kink in it, like the blade of a hockey stick,

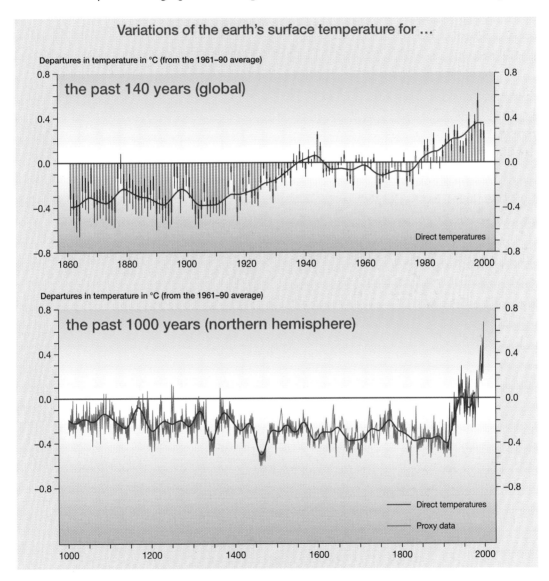

▲ The original hockey stick graph has been challenged repeatedly, but to paraphrase Mark Twain, the rumours of its death have been greatly exaggerated. Several subsequent climate reconstructions have come to the same conclusion: the latter part of the twentieth century was unusually warm.

around 1970. From that date on, the temperatures rose faster than they had for the previous nine hundred years.

It was a tricky bit of science, it relied on a lot of complex statistics, and skeptics reacted to this hockey stick in the same way as any hockey team does when threatened: they claimed the stick had an illegal curve. Actually, an *illegitimate* curve, because, they said, the statistics were wrong.

The hockey stick became a *cause célèbre*, and still is in some people's minds. Eventually, the U.S. National Research Council concluded that most of the hockey stick graph was indeed legitimate, but that the data were not good enough to be sure about the pre-1600s, nor were they finely resolved enough to conclude that the 1990s were as unusual as the hockey stick scientists claimed. But the shape of the stick was right.

Since then many other computer simulations of past climate have come up with graphs that look much the same: they swerve sharply upward around the middle of the twentieth century. These graphs, all drawn from a mix of different kinds of data, answer the question, why should we worry about climate warming? Because it's just happening faster than it should.

Why should we worry about climate warming? Because it's just happening faster than it should.

THE GREENHOUSE EFFECT

Now to close the circle: if the climate is warming unusually fast, why is that happening? One suggestion made more than a century ago was that if carbon dioxide were to increase in the atmosphere, the earth would warm. Svante Arrhenius, the Swedish chemist, calculated that a doubling of carbon dioxide would raise global temperatures about 5 degrees Celsius. (Although this was the 1890s, and he had no computer and few data, he wasn't far off current estimates.) Even before Arrhenius there had been speculation that somehow the earth's atmosphere must be like a *greenhouse*, preventing the warmth from the sun from escaping. That seemed to be the most reasonable explanation for the fact that the earth was much warmer than it should be, based on how much heat it absorbed and how much it radiated back to space.

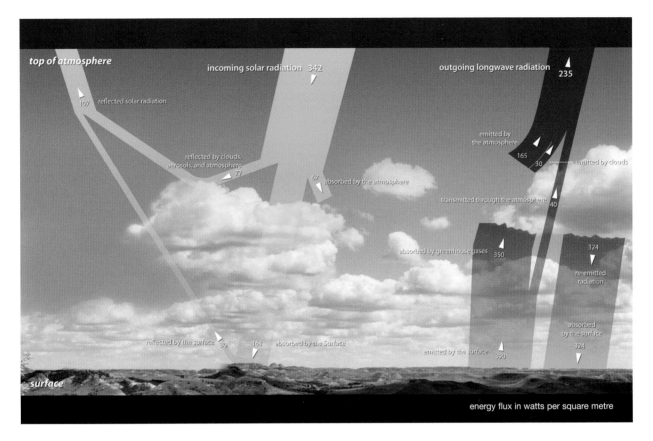

The greenhouse effect.

Even though the theory pre-existed Arrhenius, he coined the term "greenhouse effect." It's evocative but not accurate: the atmosphere doesn't actually behave like a greenhouse. The glass of the greenhouse traps warm air, preventing it from rising out of the greenhouse or being swept away by the wind. By contrast, the greenhouse gases in the atmosphere, carbon dioxide among them, trap energy being radiated back into space from the earth, then re-emit it, some out to space and some back down. By doing so, these gases boost the total amount of energy that the earth absorbs.

IT'S A GAS, GAS, GAS: CARBON DIOXIDE, METHANE, WATER VAPOUR

Carbon dioxide isn't the only greenhouse gas, but there's more of it than any other. Methane, released from belching cows, rice fields, and landfills, is more potent but much less abundant. In the decade it stays in the atmosphere, a molecule of methane absorbs twenty-five times as much infrared radiation as a molecule of carbon dioxide would in a century. But because there's so much less methane, its impact is only about a third of that of CO_2. Water vapour also has weak greenhouse capabilities, but there's a lot of it. Even the chlorofluorocarbons, chemicals that were eliminated years ago from common household products like spray cans, have an effect.

Of all the greenhouse gases, only carbon dioxide is increasing by leaps and bounds. Worldwide levels of methane have levelled off since the late 1900s, although nobody knows why. It could be a combination of things, like repairs to leaky oil and gas lines, widespread drought causing a drying up of wetlands (natural sources of methane), even a decline in rice production.

THE KEELING GRAPH

This is one of the most important graphs in the entire story of global warming. In 1958 a scientist named Charles Keeling set up instruments to measure the CO_2 in the air above the Hawaiian mountain Mauna Loa. Keeling wanted to sample the atmosphere far from any urban centres. Ever since, the Mauna Loa Observatory's CO_2 record has been the gold standard. As you can see from the graph (right), CO_2 has risen steadily. The graph has a sawtooth shape because in winter deciduous trees shed their leaves and stop taking up carbon dioxide, and the CO_2 level spikes. It then bounces back down during summer as new leaves absorb the gas.

WILD CARDS

If carbon dioxide were the only thing influencing global temperatures, there wouldn't be much debate about global warming. But it's not. Here are some of the other players.

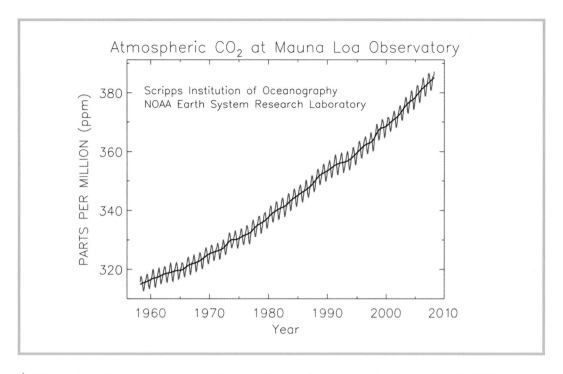

▲ This iconic graph charts the inexorable increase of atmospheric carbon dioxide since the late 1950s.

The Sun

The sun runs climate: if more solar radiation reaches the earth and stays, things get warmer. Less, and the earth cools. Some scientists who doubt the human influence argue that it's variations in the sun's output that cause the ups and downs of climate, not carbon dioxide.

Early in earth's history, the sun shone with only about three-quarters of the energy it does now. It took billions of years for it to reach its current energy output, but it's true that even now it does not shine steadily.

From the 1300s to the early 1800s, Europe experienced what's been called the Little Ice Age. It *was* cold: you couldn't get to Greenland because of the ice, in Europe canals froze solid, and glaciers oozed down the Alps into the lowlands. The curious thing is that at the same time, the number of sunspots, the small dark patches on the sun, reached record lows. Sunspots rise and fall again every eleven years or so, ranging from numbers in the hundreds

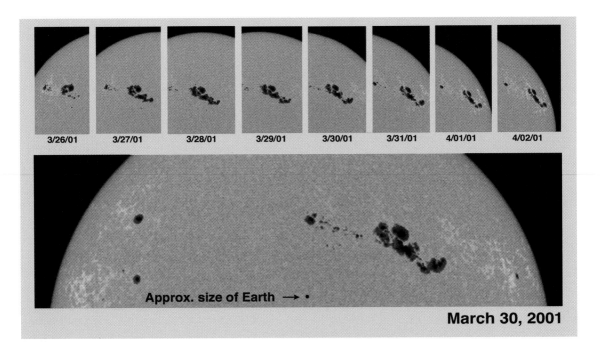

Approx. size of Earth → •

March 30, 2001

▲ Sunspots like these might become a rare sight in the next few years. The last time they practically disappeared, Europe experienced the Little Ice Age. However, this time around there is the greenhouse effect to contend with, and the result of these conflicting forces is nearly impossible to predict.

to a mere handful. Usually it's easy to count a couple of dozen spots on the sun, but from about 1645 to 1715, there were so few that the discovery of even a single one made headlines (at least among astronomers). That seventy-year period is now called the Maunder Minimum, after solar astronomer Edward Maunder.

Sunspots reached into the oddest corners of human activity. In the early 1800s, the astronomer William Herschel noted that the price of wheat fell as sunspots rose, and vice versa, because warmer weather allowed larger harvests and created surpluses. Problem is, it's not clear, even today, how the number of spots on the sun could influence climate on earth. Regardless, there seems to be a connection: the fewer the sunspots, the chillier it is on earth, or at least in some parts of the earth.

Fast forward to today: according to solar scientists, the sun, which has been extremely active over the last few cycles, is about to become very quiet, with maybe fewer spots than it's had for a century. Does this mean, as some global warming skeptics say, that we should be preparing for global *cooling?*

Hardly. The fact is that the earth's temperature over the last few decades doesn't correlate with the sunspot numbers at all. Dramatically low sunspot numbers in the near future could ease global warming for a few years, but by how much, and for how long, no one knows. And if carbon dioxide continues to build up, once the sunspots come back, warming will too.

Cosmic Rays

Some scientists have proposed that these high-energy particles from deep space might affect the climate by penetrating the atmosphere and triggering the production of clouds; the more clouds, the more sunlight is reflected back to space, and the less warming. If this is true, there could be an intriguing link to the waxing and waning of the sun mentioned above. The higher the activity of the sun, the more powerful the solar wind, which shields the earth from cosmic rays. So when the sun's activity crashes, as it did in the Maunder Minimum, you'd expect more clouds and lower temperatures.

It all seems to fit, but in fact there's no real evidence that cosmic rays actually do create more clouds. As this is being written, a large-scale experiment is in the works to test the idea.

The Earth

The earth doesn't travel in a perfect, unchanging circle around the sun. It tilts, wobbles, and veers in and out, and each of these affects the amount of energy we absorb from the sun. These effects have cycles, most of them measured in tens of thousands of years. But they have a significant impact on climate. There are three variables in total:

1. *The tilt of the earth's axis changes.* It doesn't seem like much, carrying from 21.8 to 24.4 degrees off the vertical and back again, taking forty-one thousand years to complete the cycle. As the tilt decreases, the sun becomes weaker in summer, and sometimes that's enough to allow ice to build up, especially at the poles.

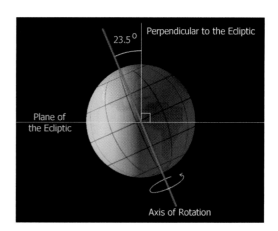

▲ The earth's tilt (23.5 degrees at the moment) changes with time, affecting climate.

2. *The shape of the earth's orbit changes.* We travel around the sun in an ellipse, an off-centre circle. These days that means that we are a little closer to the sun in January than we are in July, about 3 percent closer. But how eccentric the orbit is changes with time, again on scales of thousands of years, and increases or decreases the intensity of solar energy reaching the earth.

3. *The earth's axis wobbles, just like a spinning top, but way more slowly.* In fact, it completes just one wobble every twenty-six thousand years. Even so, the wobble is important because it influences the times of the year when we are closest to—and most distant from—the sun.

▲ Both the shape of our orbit around the sun (above) and the direction the earth's axis points (top right) change over long periods of time. Each of these orbital factors alters the climate.

Volcanoes

Volcanoes do add carbon to the atmosphere, about 10 percent of what we do, but big ones can counter that effect by emitting huge amounts of sulphates in the form of tiny droplets. Sulphates in the upper atmosphere absorb incoming sunlight and cool the earth. Mt. Agung in Bali erupted in 1963 and cooled the lower atmosphere by about 0.5 degrees Celsius; Mt. Pinatubo in 1991 rocketed more material into the stratosphere than any volcano since the immense eruption of Krakatoa in 1883. The result: global temperatures dropped by about 0.5 degrees Celsius for a year or so.

Pollution

By this I don't mean greenhouse gases, but the other stuff that we emit, including sulphur dioxide and soot. Sulphur dioxide used to pour out of smokestacks all over the world, but as industrial nations cleaned up their act and pollution controls were put in place, the amount of SO_2 and its related

▼ The eruption of Mt. Pinatubo was spectacular, but the most important effect was invisible: the deposition of huge amounts of sulphur in the upper atmosphere, cooling the earth.

sulphates began to diminish. This not only cleaned up the air we breathe, but reduced the amount of these chemicals in the upper atmosphere. There was a downside: their removal accelerated the warming of the earth. As long as these chemicals were airborne they absorbed sunlight, and scientists believe they caused the lag in warming that shows up in the records from about 1940 to 1970. As industrial activity slowed with the breakup of the Soviet Union, and as other nations cleaned up their emissions, sulphates fell, more sunlight got through, and the earth began to warm again, as it had from 1900 to 1940.

Recent research has established that the tiny particles of carbon in soot are second only to carbon dioxide in their potential for warming the earth. This stuff, called black carbon, exerts a variety of effects: it absorbs some of the solar energy being reflected from the earth, thus warming the atmosphere; it also darkens ice and snow, allowing both to absorb more heat rather than reflecting it.

Black carbon might even be the reason that glaciers in the Himalayas are shrinking so dramatically. The Tibetan side of the mountains has warmed by 1 degree Celsius since 1950, probably because air warmed by black carbon from Southeast Asia continues to wash over them. Add the darkening of glacial ice, and glacial retreat is inevitable.

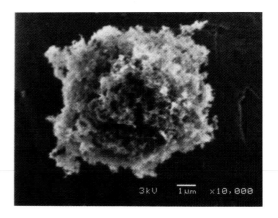

▲ A single particle of black carbon.

This is the situation we face. The scientific case for a human hand in global warming has been made, and we've moved on in our attitudes toward global warming. The question "Is it happening?" has been replaced by "How bad will it get?" and "What can we do to avoid the worst?"

Action is what's required now, and that's what this book is all about. It is a book about people inventing, dreaming, scheming, and testing—it is a highly visual telling of their fantastic stories.

Some of the people in this book, like the Barenaked Ladies and Roberta Bondar, are already famous, some—Amory Lovins, George Monbiot—are international spokespeople for the environment, but most you will be meeting for the first time. You'll find a spectacular array of technologies here as well, from the dead simple (an octogenarian's solar oven) to those that are barely at the blueprint stage. Some of the most outlandish ideas for technological solutions to the climate crisis, like sending billions of tiny solar shades into outer space, are so radical that they have split the research community. Critics fear that such a "fix" would be a disincentive for change. But if change doesn't come fast enough, if carbon dioxide emissions aren't dramatically reduced, ideas like these are just the ones that might save us.

This book is about more than just the scientists who investigate global warming, although of course they play a central role. You will also meet inventors, engineers, and even people with no particular professional specialty as they go about crafting their own response to the crisis. Some fly into the teeth of violent storms, others drill into glaciers and collect ancient ice cores. Some dream of seeding the ocean with iron powder to encourage carbon dioxide–absorbing algae blooms, others think we should mimic volcanic eruptions to cool the earth.

You'll even find people whose environmental acts have little to do with global warming, like the guys who turn high-end cars such as Hummers into hydrogen-powered vehicles, and Fergus the Forager, who scours the English countryside for his meals and never darkens the door of a supermarket. But each is part of a growing move to do something.

▼ This satellite image shows some of the glaciers in the Himalayas. The glaciers are shrinking at alarming rates, and particles of black carbon may be playing an important role in that retreat.

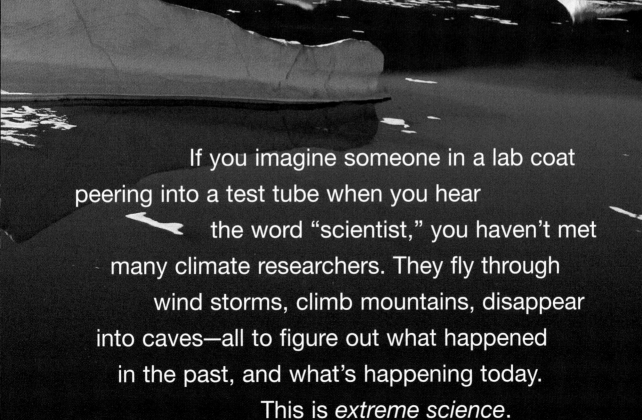

ONE

Extreme Science

If you imagine someone in a lab coat
peering into a test tube when you hear
the word "scientist," you haven't met
many climate researchers. They fly through
wind storms, climb mountains, disappear
into caves—all to figure out what happened
in the past, and what's happening today.
This is *extreme science*.

FLYING LOW

IF WE WANT TO MINIMIZE THE DAMAGE incurred by global warming, we need to know not only how much the temperature will rise, but also where the effects will be felt most. It is already apparent that the poles will warm more rapidly than mid-latitudes. Even the toughest, oldest Arctic ice, the stuff that has remained frozen for years, is being lost at an incredible rate. Gigantic ice shelves are breaking away from the Antarctic continent. But this picture is complicated by many other factors, especially winds.

Winds carry five times as much heat out of the tropics as do ocean currents. Blowing over El Niño and La Niña, the alternating patches of warm and cold water in the south Pacific Ocean, they can change the weather in Canada. Even the Gulf Stream, that powerful torrent of warm water in the Atlantic Ocean that sweeps north from the equator and keeps northern Europe temperate, depends in part on winds up to 11 kilometres high to transport its warmth landward. But those high-altitude winds are not the only important ones.

▶ Even from this altitude it's clear this is a rough and turbulent setting, but Kent Moore likes to fly much, *much* closer to where the action is.

▶ Kent Moore's idea of the perfect workplace is flying into the teeth of a gale 100 feet above a violent ocean.

If you're going to check out all the winds that play crucial roles in the changing climate, sometimes you have to get up from your lab chair and do a little extreme science. Like Kent Moore does.

Moore spends much of his time at his desk, like most university researchers. But when he gets out of the office, he really gets out. Kent Moore studies winds around Greenland.

While most climatologists see Greenland as the site of some of the thickest ice sheets in the world—a threat to sea levels, which would rise dramatically if they were all to melt—Moore looks at Greenland from the perspective of the wind. These are not just any winds: air flowing south from the Arctic whips around the southern tip of Greenland and accelerates to incredible speeds, as high as 150 kilometres per hour. And Moore flies right into them. As if that weren't enough, he likes to fly at an altitude of 100 feet!

These are critically important winds because they pull heat out of the last, dying gasps of the Gulf Stream as it completes its northern journey and its water finally cools enough to sink and begin the return trip to the tropics. But exactly what role these turbulent, high-speed winds play in climate is what Moore is after.

"It's very turbulent, it's very bumpy. The sea state is incredible. The waves are being sheared off so there's huge areas of just white foam where the wave crest has been destroyed by the wind."
KENT MOORE

▲ Ferocious winds off the coast of Greenland soak up the last vestiges of warmth from the Gulf Stream.

This is when climate research becomes high adventure. To study these winds, Moore and his fellow researchers climb into a specially outfitted BAe 146, a British plane flown by Royal Navy pilots experienced with low-level flight in turbulent conditions. The plane is loaded with instrumentation so the winds can be characterized in detail as the plane flies through them.

The turbulence is rough all right, but it's exactly that turbulence that is transmitting heat and moisture from the ocean to the atmosphere. On this day, however, the conditions are too much, even for this crew and this airplane. The pilots can't see far enough ahead of them, so they have to climb to a higher altitude, about 2000 feet. But all is not lost—they're still able to release dropsondes from the plane, small, paper towel tube–sized devices on small parachutes, each carrying its own barometer, thermometer, GPS, and humidity sensor. The dropsondes report back to the plane as they descend.

From the dropsondes, Moore can tell that the winds are much more intense near the ocean surface than at the altitude to which the plane has retreated, and that finding in itself is significant: "This shows you the influence that Greenland has. Greenland was channelling all that energy down to these very, very strong winds right down near the surface. It doesn't happen very often that you get stronger winds at the surface than you do at higher altitudes."

Moore is far from finished his study of

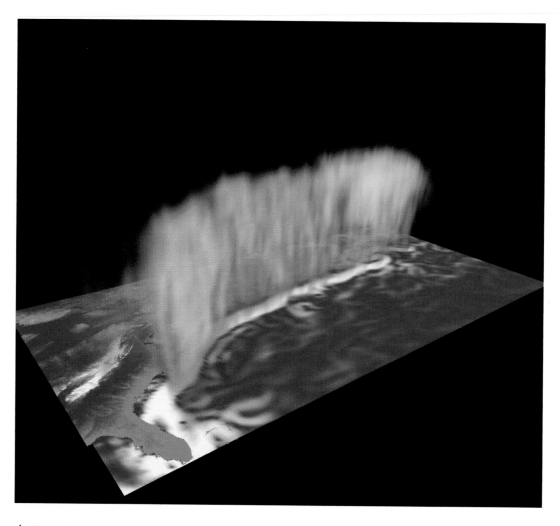

▲ Extending from the Florida coast, the Gulf Stream brings warm water to the north. But recent studies suggest that it's not so much the water, but rather a high-altitude corridor of warm air (seen here as orange and yellow) that is the prime mover behind the Gulf Stream's benign influence on European climate.

the powerful winds around Greenland. He is confident that they play an important role in determining climate in the North Atlantic, but the devil is in the details. Flights like today's, where he didn't get to fly exactly where he wanted to, are frustrating because, as he says, "we need to make some serious decisions about how we're going to modify our use of carbon and we can't delay doing that. We need to do it now so that we can, in ten years' time, have better predictions as to what's happening in the climate system."

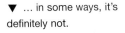
◀ In some ways, it's a typical lab …

▼ … in some ways, it's definitely not.

The Natural World Under the Ice

Perhaps this should be called the "Unnatural World." Over the last few years, scientists and citizens have watched with fascination and disbelief as huge shelves of ice have broken away from the Antarctic continent and crumbled. Satellite images reveal what might not otherwise have been seen or recorded: the most recent piece, the Wilkins ice shelf, broke away at the end of February 2008. Dr. David Vaughn of the British Antarctic Survey had predicted this event back in 1993, but he figured it would take thirty years. It took half that time.

An even more spectacular event happened in March 2002, when a 200-metre-thick piece the size of Prince Edward Island disintegrated into the ocean. The Larsen B ice shelf was twelve thousand years old.

Antarctica is warming, and the breakup of these ice shelves is one of the results. Because they are shelves, already floating, they do not lead to significant rises in sea level, but they are a cautionary sign none-theless. One research group has taken an unusual tack in response to the Antarctic ice shelf destruction: they are looking at what was living under the shelves before they broke up, and they are turning up some dramatic surprises.

After the breakup of the Larsen B sheet, these scientists, some fifty of them, launched the Polarstern research expedition. Why? Julian Gutt is chief scientist of Polarstern, and in his opinion, "Life under ice shelves is maybe the least known ecosystem—marine ecosystem—on earth."

The scientists sent nets, buckets, and remote-controlled submarines to the ocean floor to find out what had been living under more than seventy storeys of ice—living there, sealed off from the outside world, for thousands of years. At depths of 200 metres, they found an array of organisms that would normally be found in the deep ocean, 2000 or even 3000 metres below the surface. Finding food under the incredible thickness of an ice shelf is probably like finding food in the deep ocean—difficult. Only those creatures that can survive the relative scarcity of the deep ocean cold colonize the seabed under the shelf. But things have changed dramatically since the ice shelf fell apart in 2002. The Polarstern scientists found that new species had already started to move in: "We counted quite a high number of juvenile sponges. These juveniles indicate a shift in species composition after the collapse of the ice shelves."

Soon, at least from a biological perspective, the sea floor examined by the Polarstern expedition will betray no trace that it was once in the dark and cold of the Larsen B ice shelf's shadow.

▲ Sea anemones now flourish in the waters formerly shadowed by the Larsen B ice shelf.

▲ Life on the sea floor began to change as soon as the ice shelf disintegrated.

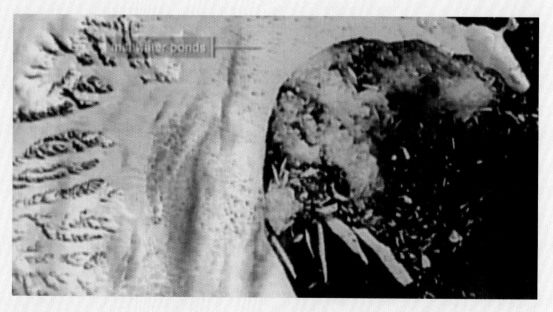

▲ The catastrophic collapse of the Larsen B ice shelf is recorded here by the MODIS satellite.

▲ These fast-growing sea squirts are the marine equivalent of "early adopters." They move in when an ice shelf collapses because the undersea environment changes dramatically.

◄ Ice at the edges of Antarctica floats on the ocean.

DEEP SNOW

GLACIERS HAVE ALWAYS BEEN POPULAR with tourists, but today they are attracting the attention of more scientists than ever before. With good reason: a retreating glacier is a sign that the climate is warming and more ice is melting than is being added. Glacial advance and retreat in the past marked the ebb and flow of earth's major ice ages. During the earth's coldest epochs, glaciers buried vast areas: most of Canada was completely icebound fifteen thousand years ago. The ice was thick too, so thick (3 kilometres deep in some places) that the land is *still* rebounding from the great weight that was removed when the glaciers melted. That rebound is so slow—about a centimetre a year and declining all the time—that it's expected to continue for another ten thousand years, almost as long as it's been going on.

▶ *Early Morning, Rocky Mountains,* by J.E.H. MacDonald, 1926.

▲ Glaciers advance and retreat, but North America has never been without glaciers since the first people arrived.

How glaciers wax and wane is pretty straightforward: when it's cold, snow falls on the glacier, adding to it. The glacier flows downhill and along valleys. When it's warm, the glacier starts to melt, especially at the toe, and begins to recede. Actually, it's both advancing and receding at the same time, but if warming prevails, the entire glacier will shrink back. That, of course, is what we've seen happen all over the globe in the last few decades. How much of that shrinkage is due to human-induced global warming is uncertain—some glaciers started shrinking before industrial activity likely

▼ As this graph from the World Glacier Monitoring Service makes clear, glaciers all over the planet are in full retreat.

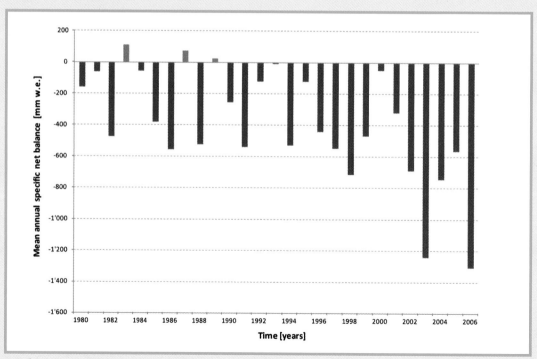

exerted a major effect—but there's no question humans have caused a substantial part of it. The World Glacier Monitoring Service put it this way in the 2007 Glacier Mass Balance Bulletin: "Continued shrinking of glaciers and ice caps may have become primarily forced by human impacts on the atmosphere."

If it's true, as some skeptics argue, that a number of glaciers in the world are advancing (and some claim that more than half are), it's hard to conclude that from the latest graphs available from the World Glacier Monitoring Service.

A stone tool is useless to an archaeologist if she doesn't know how old it is or where it was found. A snowfall is the same.

▼ In this graph the downward trend is more obvious. The units on the graph—mm w.e.—stand for millimetres of water equivalent.

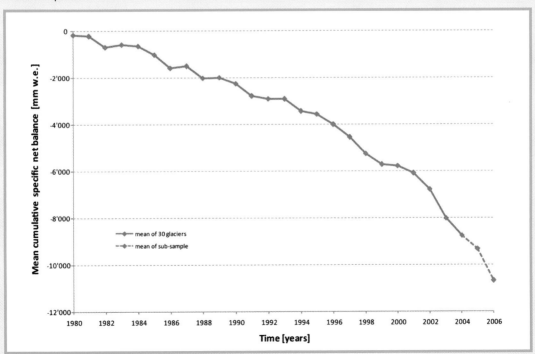

The number of glaciers losing ice and the rate at which they're losing it are rising all the time. But even a hundred years ago the same story was beginning to unfold. The Illecillewaet glacier spills out of the mountains close to British Columbia's Rogers Pass, named for Major A.B. Rogers, who explored the area in the early 1880s while seeking a railway route through the Selkirk Mountains. Within a few years the Canadian Pacific Railway had built the Glacier House hotel, no more than a half-hour walk from the tip of the glacier. The combination of the railroad and that short walk made Illecillewaet the most visited glacier in North America. Since that time the glacier has receded at least 800 metres (despite advancing for a short time in the 1970s).

Kate Sinclair is a Ph.D. student at the University of Calgary. Glaciers are her thing. She is trying to figure out the details of how glaciers add new ice (even if overall they're in retreat). She wants to know how much snow is added in winter, and more impor-

tantly, where that snow comes from and how it gets there. You might think snow is snow, but in fact it contains chemical clues to its origin. As Sinclair says, "What we're looking at specifically is the oxygen and hydrogen content of the snow, and it changes depending on the storm pathway, how far the storm has to travel."

Both oxygen and hydrogen atoms exist in different forms, called isotopes, with slightly different weights. These differences in weight, or mass, are crucial. For instance, heavier isotopes rain out first, meaning that the farther an air mass travels, and the more moisture it loses, the fewer of these heavy isotopes are left. Temperature also affects the mix. Put it all together, and the chemistry writes a history of the storm. "By analyzing the isotopes," Sinclair explains, "we're actually able to pick up very distinct differences in the snowpack. It really shows us the storm trajectories."

Sinclair samples the most recent layers of snow, takes them back to lab, and analyzes

▲ Just another day ...

▲ ... at the office.

them. By matching the isotopes with the weather patterns that existed when the storms occurred, she is able to identify where they came from: sometimes Alaska, sometimes the south. This kind of information provides another skin to the onion: if a particular snowfall came from Alaska, and happened at a time when global patterns of air movement had moved the jet stream northward, then an important piece of information about the origin of snowfall in the eastern Rockies has been put in place.

Sinclair's research is all about provenance, the identification of the origins of things. A stone tool is useless to an archaeologist if she doesn't know how old it is or where it was found. A snowfall is the same.

But this is only the beginning. The next step is to take this storm-tracking ability deeper into the past, by trying to find similar storm signatures in ice cores hundreds or thousands of years old. If that's possible, we will have a much more detailed look at how glaciers are built.

▼ Dan McCarthy of Brock University has monitored the Illecillewaet glacier in British Columbia for years. Here you can see just what's happened over the last century. A hundred years ago the glacier extended beyond the borders of this picture; by the 1930s it was partway up the slope; and today it ends even further uphill. The dotted line across the top of the ridge suggests how thick Illecillewaet was in the 1890s.

Americans work their way through something like a hundred billion plastic bags every year.

The Ocean as Landfill

Ian Connacher

It's a tangled web, this climate change stuff. It bleeds into every other environmental issue: habitat loss, biodiversity, pollution, spread of disease, and the price of oil, to name a few. Climate change is the overriding issue of our time, but even if it weren't, anyone concerned about the environment would inevitably bump up against it.

In his recent film *Addicted to Plastic* (slated for broadcast in several countries), Toronto filmmaker Ian Connacher records his journey around the world sizing up the problem of discarded plastic. What's the link between plastic and global warming? You need oil to make plastic, and a lot of plastic is being made. For instance, Americans work their way through something like a hundred billion plastic bags every year, sucking up twelve million barrels of oil along the way. That's reason enough to be concerned about our reliance on plastic. But this is not so much an issue of oil consumption or emission of greenhouse gases as of attitudes and habits.

▲ There's no trouble finding plastic in the North Pacific Central Gyre.

Take this example: once Connacher had his epiphany ("When I woke up and realized just how much plastic I touched in a day, I knew I had to do something"), he hitched a ride on a boat travelling to the North Pacific Central Gyre, a vast area of the Pacific Ocean where the atmosphere presses down on the water, actually depressing it, and allowing garbage to swirl inward and concentrate toward the middle. But the "middle" is an area the size of western Europe, and while it is *full* of plastic, Connacher found it might not look that way at first glance.

"In my mind I had an image of a floating landfill out here, but it's not that. It's a chunk here, a piece there, but when you do it for an hour, and realize just how small the section is that you are able to search, and the fact that only half of all plastics float, you begin to get a sense of just how much plastic must be out here."

The gyre is home to plastic in its myriad forms: bottles, nets, light switch covers, even fluorescent tubes. Eighty percent of this floating garbage comes from the land; in some areas of the ocean, the ratio by weight of plastic to phytoplankton, the plants of the sea surface, is ten to one. All this plastic has been spilled or thrown away. And, Connacher notes, "Every piece that's ever been made, except for a small amount that has been incinerated, still exists."

Connacher does find room for optimism: scientists working on commercially viable biodegradable plastics; entrepreneurs turning waste plastic into new products (as we'll see on page 184, Interface Carpets would love to start mining landfills, arguing that they are the oil wells of the future). Even so, the overwhelming images in his film are of seas, hills, beaches, and mountains of plastic, evidence of our ignorance of, or carelessness about, exactly what we're doing. In the end, once you have the knowledge, it's all about attitude. And that's what connects plastic to all other environmental issues, including global warming. As Peter Applebome put it simply and directly in *The New York Times*, "If we can't change our behaviour to deal with this one, we can't change our behaviour to deal with anything."

▲ Signs of the present …

▼ … and of the future.

WILL THAT BE ON THE ROCKS, OR STRAIGHT UP?

THE NATIONAL ICE CORE LAB IN DENVER, COLORADO, HOLDS SOME amazing treasures. The lab seems secure enough, but if the temperature inside were to change drastically, those treasures would melt away into worthlessness. According to Todd Hinkley, "It's been estimated that if this entire collection were to run out into the parking lot, it would cost sixty or eighty million dollars to replace it."

The lab keeps its ice cores—long tubes of ice about the diameter of the fat end of a baseball bat—at minus 36 degrees Celsius. There are about 15 kilometres' worth in this lab, and they are one of the most valuable sources of information about past climate you can get. When snow falls in places like Greenland and Antarctica, it may never melt. Instead, it is buried by the snow that falls on top of it, and eventually those snow crystals are pressed into ice. If you're careful to find the right place, the ice pack may record hundreds of thousands of years of snow accumulation. And if you know how to read those records, they reveal a stunning amount of information.

"If you look at all the coal, oil, and natural gas that has been burned on this planet ever by humans, 50 percent of that was burned since the early 1970s."
TODD HINKLEY

Am I making this sound easy? That would be a huge mistake on my part. To retrieve that information you have to go to some of the most inhospitable places on earth, places where the record of extremely cold weather remains pristine. Then you have set up the drill equipment and settle in for the long haul. The months-long, cramped-quarters, we're-out-here-in-the-middle-of-nowhere-and-it's-cold haul. Drilling into the ice and pulling out cores. The farther down you drill, the older the core you get. You have to drill down kilometres. So far the record is a core from the Antarctic that goes down more than 3 kilometres and back eight hundred thousand years!

◀ Unremarkable in appearance, this warehouse of ice cores holds the secrets to tens of thousands of years of earth's climate.

▲ To reach deep into the past, you have to go to the ends of the earth.

A GUIDE TO ICE CORE READING

1. *Summer snow freezes white, but winter snow freezes clear.* This is because the chemistry and texture of the snow differs from season to season. While it's true that at a glance you can tell which season you're looking at, it gets trickier the deeper you go: the farther down the ice, the more it's compacted by the weight of the ice above it, and the harder it is to read the core. Of course, the thickness of the band tells you how much snow fell that season, or that year.

2. *Air bubbles are trapped in the ice as it forms—and they never escape.* So if you have a core from one hundred thousand years ago, the bubbles in it contain a sample of the atmosphere from one hundred thousand years ago. How much carbon dioxide was in the air then? How much methane? The answers are in these cores. As Todd Hinkley says, "Ten percent of the ice is the atmosphere. So we can crush that ice and remove carbon dioxide, methane, other gases and tell, for example, twenty thousand years ago, how much CO_2 was in the atmosphere."

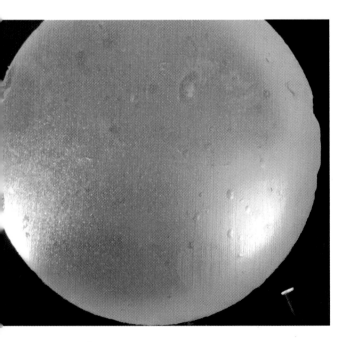

▲ The faint bubbles in this ice core are fossils of the earth's atmosphere.

3. *Ice cores can also tell you how warm it was—or wasn't.* This is where the chemists come in: the earth's atmosphere contains various versions of oxygen, and they all have slightly different weights. The two most important are the most common one, oxygen-16, and the much rarer, and heavier, oxygen-18. The warmer the ocean, the more readily the heavier one will evaporate, then fall back to earth in snow. So the more oxygen-18 you find in ice cores, the warmer it must have been when that snow fell.

4. *That is not all ice cores can do.* If all ice cores did was tell us what the temperature and the atmosphere of the ancient earth were like, that would be fantastic enough. But they don't stop there. Traces of a variety of materials are also caught up in the falling snow and deposited. So, for instance, Hinkley has shown that you can tell from the bits of heavy metals in the core, like lead and indium, not only when volcanoes were erupting, but also what kind they were. A few thousand years ago, the amount of indium falling on Antarctica decreased dramatically, suggesting that there was a switch in eruptions from mainly "black rock" volcanoes, those most often found on oceanic islands, to "white rock" volcanoes, which are generally found inland.

This switch happened about six thousand years ago, possibly as a delayed response to the melting of the great ice sheets of the last glaciation. Higher sea levels might have suppressed the island volcanoes, while the melting back of the ice sheets could have unweighted the landlocked volcanoes. It's a guess at this point, but speculation such as this illustrates brilliantly how rich in detail the ice cores can be. A thousand years from now, if scientists are still drilling ice cores, they'll see a dramatic drop in atmospheric lead from the 1980s, when many countries switched to unleaded gasoline.

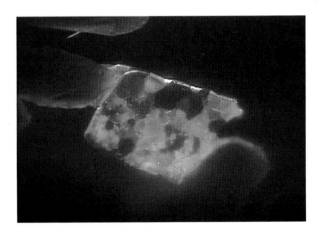

▲ The job of recovering ancient gases from the ice cores requires painstaking care.

CAVE WOMAN

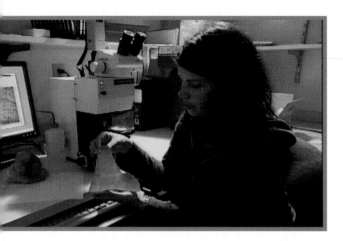

"Stalagmites have the advantage of being easily datable in calendar years back to several hundreds of thousands of years."
JESSICA OSTER

WE CAN NEITHER ANTICIPATE WHAT the climate might do in the future nor even evaluate the extent of our role in climate change if we don't know how the climate has behaved in the past. No single record of past climate, whether from tree rings, ice cores, or ancient pollen grains, is enough. Each has its limitations, and any new way of calibrating the past is welcome. That is what takes Jessica Oster deep into caves in the foothills of the Sierra Nevada in California. She's looking for a perfect example of one of those spooky growths called stalagmites that rise from the cave floor.

Oster explains, "There are very few sources of climate records from low-latitude terrestrial sources. Stalagmites have the advantage of being easily datable in calendar years back to several hundreds of thousands of years."

Amazing but true. Stalagmites form at an achingly slow pace as mineral-rich water drips onto the cave floor, then evaporates and leaves the minerals behind. Because stalagmites contain trace amounts of uranium, which decays at a slow but precisely regular pace over time, events can be dated back several hundreds of thousands of years, according to Oster. As long as the drip continues to come from the same place (and caves are remarkably constant environments), the stalagmite will grow upwards, and may reach several metres in height. Stalagmites have growth rings, like trees, reflecting variations in their deposition with the seasons. The water that drips down to form them comes from the atmosphere (and so may contain chemical traces of temperature and carbon dioxide) but also mixes with water in the soil. Soil water has its own carbon dioxide content, so you can see that analyzing the climate signal present in a stalagmite is not straightforward and simple.

But the challenge of deciphering the signal is well worth it. Some stunning records have been read from stalagmites. Scientists in China have created a 224,000-year history of the monsoon rains and have shown that they vary with the regular tilts and wobbles of the earth as it orbits the sun. And in California, Jessica Oster is hoping to create her own record of the deep past.

She is working in the southern Sierra, where it's possible that climate signals go back more than a million years. But before she can try to read those signals, she has to get the right stalagmite. And that is what she was doing when *Daily Planet* was there.

Together with Steven Fairchild, who's with the company that owns this cave, she was looking for a candidate stalagmite for analysis. And one in particular appealed to

▲ A century is no more than a short segment of this stalagmite.

▲ The effects of global warming are expected to be dramatic in high-altitude areas such as the Sierras.

"In future climate change with global warming, areas of high altitude are predicted to see changes a lot more quickly and a lot more dramatically than areas of lower altitude."

JESSICA OSTER

her: "So here you can see that this is the main drip centre that the stalagmite grew from. But there's also this back here, this extra lump indicating that either the drip moved from here to here, or vice versa. So either of these might be the younger one. But having the two means that I have essentially almost double the climate record."

It might be double the climate record, but as Oster saws through the stalagmite, she finds it's not perfect: "I can see there's some rings in here, some sort of brown rings that suggest to me that these could be times when it stopped growing or the growth slowed down … it means there's a large gap in time represented there."

This stalagmite will be taken back to the lab to remove some samples, but in the end, it's going to come right back here. Oster plans to use laser scanning and 3-D printing to make an exact replica that she can continue to study. Then she'll bring the stalagmite back and glue it onto its stump. Who knows? It might just pick up where it left off, recording what's going on in the atmosphere outside this cave.

"We're in the foothills of the Sierras, and in future climate change with global warming, areas of high altitude are predicted to see changes a lot more quickly and a lot more dramatically than areas of lower altitude. So having these past climate records from these high-altitude areas could show us how these areas responded to climate change in the past."

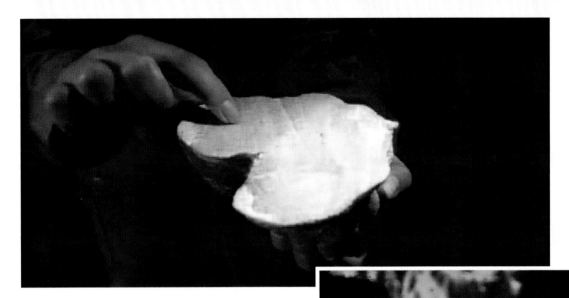

▲ Cross-sectioning the stalagmite reveals the details of its history.

▶ Jessica Oster finds what could be the perfect stalagmite for reconstructing the climate in the southern Sierra.

▼ Stalagmites form at an achingly slow pace as mineral-rich water drips onto the cave floor.

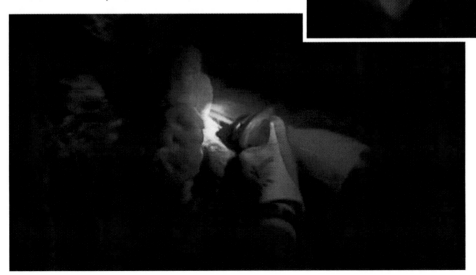

Urchins

The Polarstern expedition (see page 8) has revealed what life is like under the Antarctic ice shelves—before global warming triggers the collapse of those shelves. However, most scientific interest in life in the world's oceans is focused on those organisms that might suffer as the world warms.

Dr. Gretchen Hofmann is a marine biologist at the University of California, Santa Barbara. She studies sea urchins, spiny relatives of starfish that are abundant throughout the world's oceans. They are best known to most people as the source of *uni*, a sushi favourite crafted from the urchin's reproductive organs. But sea urchins may soon be taking on a much more significant role. Dr. Hofmann explains that although urchins "aren't as charismatic as penguins or polar bears, they really have a lot to say about what happens in the ocean because they live in the water, and their biology and how they work can instantly tell us what impact the chemistry in the water is having."

Sea urchins begin life as free-swimming larvae, which, although tiny, have a simple skeleton. Eventually, the larvae settle to the ocean floor and construct a more elaborate, shell-like version of their skeleton. There are worries that as carbon dioxide continues to accumulate in the atmosphere, more and more will dissolve in the world's oceans, and the acidity of the water will increase. It's estimated that the oceans have absorbed about 40 percent of the carbon dioxide emitted by humans over the last two centuries, and that the rise in acidity has already begun. The more acid the waters, the more difficult it might be for urchins and a variety of other animals (corals, mussels, and scallops among them) to make those skeletons. And in a globally warmed world, that will not be the only challenge for these creatures. "We are facing a double jeopardy situation in the world's oceans," Dr. Hofmann notes. "We know that they will acidify and they will also become warmer, so in the

future, high CO_2 world, we face a warming and acidifying sea."

Dr. Hofmann and the researchers in her lab are trying to estimate just how much urchins might be affected in warmer, more acidic oceans. So they fertilize urchin eggs and watch the larvae develop under those conditions. The effects can be subtle, so subtle that they have to photograph and carefully measure every larva. But when they do so, what they've seen so far suggests that these animals are having a rough time.

"When you raise sea urchin larvae at levels of CO_2 that we predict for the future, they are able to make their skeleton, but it's not normal. They are shorter and stumpier, and that's going to affect how they swim and move around in the water. We also know that they are doing that by turning up the volume on some of their genes that are involved in making that skeleton.

"But most interestingly, making that skeleton, although they are doing it under these abnormal conditions, comes at a

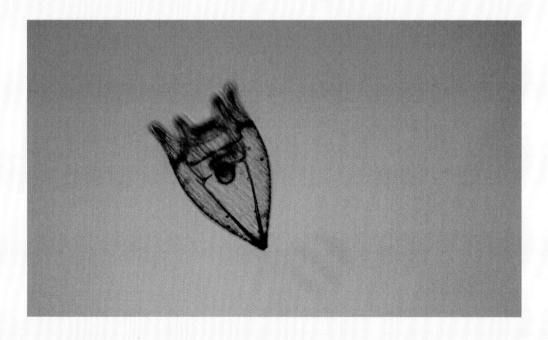

high cost. We found that larvae that develop at high CO_2 conditions are more sensitive to temperature. They are less able to cope with high temperature."

Even if the urchins are able to compensate for warmer, more acidic waters, the fact that their skeletal structure is compromised is bound to have adverse effects on their survival. For one thing, larvae that are less mobile will inevitably settle to the sea floor close to where they hatched. If that habitat becomes degraded, their chances of survival are reduced compared to organisms that move great distances before settling down. Of course, moving great distances through the ocean also raises the chance that larvae will die before they ever become adults. But that is a threat they have evolved to cope with; oceans that threaten to warm and acidify at a rate never seen before are another story.

◄ ▲ Sea urchin larvae like these may be especially vulnerable to warmer and more acidic oceans.

"We are facing a double jeopardy situation in the world's oceans."
DR. GRETCHEN HOFMANN

THE ICEMAN
COMETH ... SLOWLY

YOU'D THINK IF YOU WERE GOING TO RESEARCH glaciers, you'd want to be in Iceland. How much easier could it be? The country is covered with them. In some places you're barely out the door and there's a glacier in front of you.

One of them, Vatnajökull, is the biggest glacier in Europe. Its awesome 8100 square kilometres could completely cover Prince Edward Island with plenty left over. In some places Vatnajökull is a kilometre thick. The second-biggest glacier in Iceland, the "long glacier" Langjökull, is just under 1000 square kilometres. And that is the one that interests Finnur Pálsson.

He is an Icelandic scientist who wants to find out exactly what's happening to the glaciers in Iceland as the world warms. An easy assignment, right? Well, no. Here's a typical research day for Pálsson.

"Getting to the glacier can be very easy if you're lucky but it can also be hazardous."

Land that has
been covered
with ice for
nearly five
hundred years
is being
exposed.

Ah, those Icelanders and their penchant for understatement. There's a whiteout today. Pálsson couldn't go anywhere if he didn't have a GPS, but even with that, he's having a tough time. For one thing, the snow he's driving through is getting deeper and deeper. He can release some of the air from the tires, so they'll sink deeper into the snow and grip better, but he can't keep doing that forever. After a while the team is moving just a few metres at a time, but they're close enough to their destination for Pálsson to do his thing. First, the ground-penetrating radar: "To measure the thickness of the ice we use ground-penetrating radar, which transmits radio waves through the ice to the bedrock where they're reflected, and from that we can calculate the thickness of the ice at each point."

Then, a steam drill creates a hole down which a wire is threaded. It will sit there until fall to measure how much snow melts during the Icelandic summer. Digging another hole reveals that the snowpack at this place near the edge of the glacier is about a metre and a half thick. That's a lot of snow, but probably only about a quarter of what's on top of the glacier. But even with this much new snow, Langjökull is shrinking.

◀ The glacier seems impregnable, but the measurements show otherwise.

▲ Digging another hole reveals that the snowpack at this place near the edge of the glacier is about a metre and a half thick.

"We've been measuring the mass balance of the glacier now for a period of twelve years, and for the whole of this period the glacier has been getting smaller. It loses approximately 1 metre of mass every year."

That 1 metre of ice is *the amount by which the glacier thins, across its entire breadth, every year.*

"This shows us that the glacier will disappear almost totally within 150 years. And if it gets warmer, as is predicted, it will disappear even faster than that."

The glaciers of Iceland reached their peak in the late nineteenth century. But since then they've been in retreat. In 2007 every glacier in Iceland retreated except one, some close to 100 metres. Iceland's glaciers have melted faster for the past ten years compared with the 1930s and 1940s, when temperatures were also warm. Currently, land that has been covered with ice for nearly five hundred years is being exposed. Pálsson's caught: needing to move fast, forced to move slowly.

THINK DEEP

Iceland might be the most environmentally friendly country on earth.

ICELAND IS UNIQUE: IT HAS STARKLY beautiful scenery, it has Björk, and it might be the most environmentally friendly country on earth. A certain amount of that greenness is luck: the country sits astride the Mid-Atlantic Ridge, where magma from deep in the earth flows to the surface between the separation of two tectonic plates. Sometimes the result is overt volcanic activity, but even in the absence of eruptions, heat is close to the surface, and that means geo-thermal power.

Iceland has five geothermal power plants, and to hear what each provides is bound to make those of us who live in countries less blessed green … but this time with envy. Albert Albertsson runs one of those five power plants: "The water comes from a depth of approximately 2000 metres and the temperature is about 240 degrees Celsius. In the ground it's water under high pressure. When we open the reservoir, the fluid starts to boil, and that drives the fluid up to the surface."

What comes up is a mix of steam and liquid. The steam goes directly to an electrical generator, where it produces 80 megawatts of energy, enough to power a small city. The energy is

then used to heat groundwater to 115 degrees Celsius, at which point the water is distributed to homes. And the whole system is virtually nonpolluting. Geothermal energy like this makes up about 25 percent of Iceland's total energy needs (fossil fuels make up less than 1 percent). But are Icelanders content with that? Well, no. They are now contemplating something much more spectacular called the Iceland Deep Drilling Project.

Geologist Gudmundur Fridleifsson is a perfect example of a scientist who, faced with something surprising and unknown, decides to pursue it. In the end, he might make it possible to generate geothermal energy on a scale never seen before. Twenty years ago he made a fluke discovery on a drilling project—he suddenly struck something very hot: "It just simply sort of went off record and out of limits. So we didn't know if it were 400 or 500 degrees or whatever. But it just went off the scale."

▲ In Iceland the heat of the earth's interior is always close at hand.

He had tapped into supercritical fluid, water that is both extremely hot and so pressurized that it behaves like a hybrid of a gas and a liquid. This wasn't just ordinary geothermal-type heat. This was something different, something that could provide, according to computer models, about ten times as much power from a single well as any other geothermal project. But the challenges are on roughly the same scale as the potential rewards. For one thing, the well has to be something like 5 kilometres deep, twice as deep as any other geothermal well in Iceland.

After choosing a site, right where the Mid-Atlantic Ridge runs into land, Fridleifsson and his team started drilling in 2004. But it turned out to be a risky business: "During a flow test the well collapsed at a depth of about 3000 metres. And they tried to clean it, but it turned out they couldn't, so we had to abandon the well and select a new one."

▲ Geothermal serves 25 percent of Iceland's energy needs.

Choosing a new site took another two years, and now they're drilling again. Even if well collapse is prevented the next time, there is still an element of danger here. Nobody is absolutely sure how the supercritical fluid will behave once they tap into it. And there's always the small possibility they will hit magma, which would qualify them as creators of the world's first man-made volcano.

If the Deep Drilling Project works, Iceland will move from being the world's leader in geothermal energy self-sufficiency to the world's leader in exporting energy generated from geothermal. The whole project will likely take ten or fifteen years, but Fridleifsson and his co-workers envision a day when Deep Drilling energy could be used to make hydrogen for export. Or, the same drilling techniques could be adapted to tap geothermal energy from the hot ocean vents, where supercritical fluid is much more likely to be near the surface (even if that surface is under hundreds or even thousands of metres of ocean). All of that would be even more mind-boggling than a Björk album.

▲ Geothermal is already big business in Iceland. But the Deep Drilling Project is something else entirely.

▼ ▲ Iceland's enormous capacity for geothermal may one day make it a significant exporter of energy in the form of hydrogen.

One-Twentieth and Seven

These are the key numbers when it comes to Greenland. The island contains one-twentieth of the world's ice, and if all that ice melted, sea levels would rise 7 metres. That would pretty much inundate many of the world's largest cities, including New York, Shanghai, and Beijing.

Over four summers between 2003 and 2007, Greenland lost an average of fifty billion tonnes *more* ice than it gained in snow every winter. Jakobshavn Isbrae, the largest of Greenland's outlet glaciers—ice that flows down the fjords to the ocean—accelerated dramatically in 2007. In a period of about ten years from the early 1990s to the early 2000s, this glacier more than doubled its speed, from a little less than 6 kilometres a year to more than 12. It's not just the rate of flow that concerns researchers. During the summer, vast, shallow lakes of melt-water form on the surfaces of glaciers. It's suspected that this water seeks and finds channels through which it flows from the

▶ At the edge of Greenland's melting glaciers, vast lakes form and disappear, sometimes in a matter of days.

surface down into the depths of the glaciers. Scientists have seen huge waterfalls form and then disappear in the space of a day or two. One lake 4 kilometres long and 8 metres deep vanished into the ice in just an hour and a half! The crucial question is whether this fast-flowing water is lubricating the glaciers, accelerating their flow. Many suspect this is true, although scientists believe this acceleration may not be as dramatic as they once thought, at least for the outlet glaciers.

As yet, there are not enough data to confirm this hypothesis.

One of the most puzzling aspects of Greenland is that 120,000 years ago, temperatures here were 50 degrees Celsius higher than they are today, and yet sea levels were only about a metre or two higher than they are now. And there was still ice on Greenland. Much warmer, yet still lots of ice—it doesn't seem to make sense.

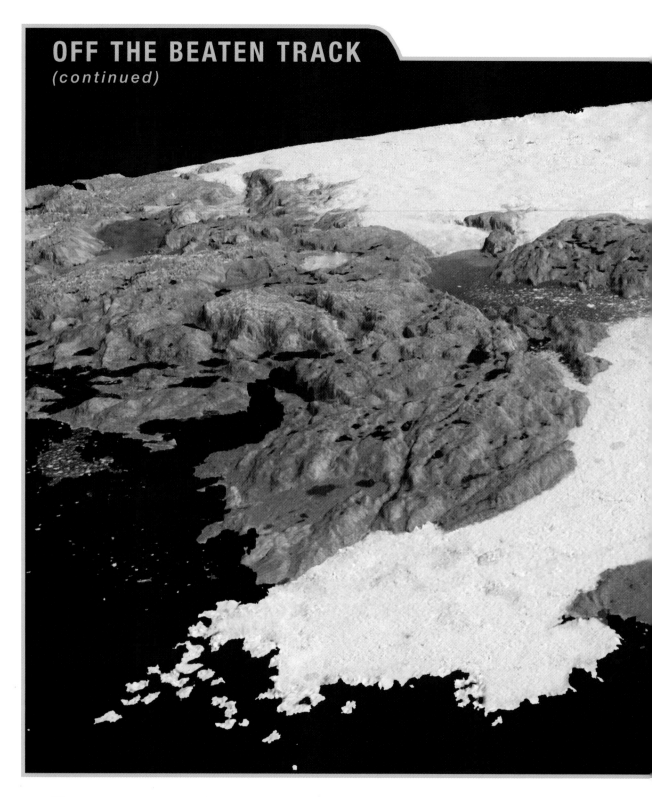

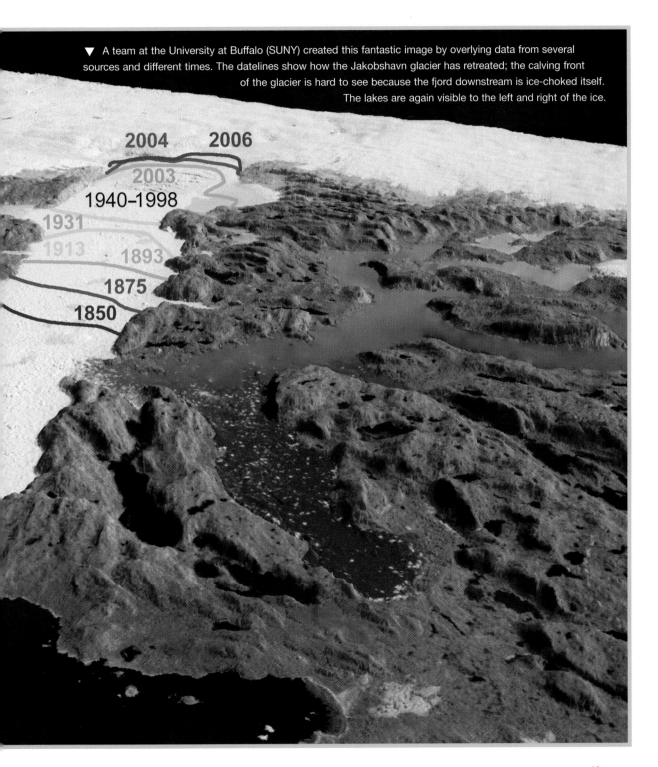

▼ A team at the University at Buffalo (SUNY) created this fantastic image by overlying data from several sources and different times. The datelines show how the Jakobshavn glacier has retreated; the calving front of the glacier is hard to see because the fjord downstream is ice-choked itself. The lakes are again visible to the left and right of the ice.

TWO
Desperate Measures

We've already begun to feel the effects of
global warming. The question is, what we can do about it?
Scientists, inventors, and engineers are coming up
with their own techno-fixes for climate change,
and while some of their ideas may seem far-fetched,
it's all about taking action.

THINKING THE
UNTHINKABLE

THE SCIENTIFIC EVIDENCE IS inarguable: we are warming the planet, we're warming it faster than ever before, and, most important, we really don't know what the results will be. Only the most naively optimistic—or those in complete denial—would argue that things will get better if we keep pumping carbon dioxide into the atmosphere.

So what can we do? If we believe the dire predictions of Sir James Lovelock (see page 60) or George Monbiot (see page 112), it's either already too late or the steps we have to take are so staggering as to be paralyzing. On the other hand, if there were enough time, we could cut carbon emissions gradually, ease the accumulation of CO_2 in the atmosphere, and bring the whole runaway climate engine under control. But it's become clear that we likely don't have the luxury of that kind of time.

The urgency of the situation has prompted climatologists and engineers to discuss measures that even a few years ago would have been dismissed as unthinkable. These "techno-fixes" look at engineering ways of cooling the planet instead of reducing emissions. But this means we would be tampering with the climate in two ways at the same time—warming it with carbon dioxide, and cooling it with so-called geo-engineering.

Undoubtedly, some of these schemes are ingenious, but they carry significant risks:

- If it's possible to cool the earth, people will think the problem is solved and will have little incentive to curb their carbon emissions.

- If we begin to employ a scheme to cool the planet, and carbon emissions are unchecked, the cooling scheme can

never be stopped. One climate modeller, Raymond Pierrehumbert of the University of Chicago, envisions a future where the amount of CO_2 in the atmosphere has quadrupled, but temperatures have been held in check by geo-engineering. If political upheaval intervenes and the geo-engineering is halted, the earth's temperature in the tropics could rise 7 degrees Celsius in three decades, with unimaginable impacts.

• Inevitably, there will be negative side effects. Some we can anticipate, like the gradual acidifying of the ocean by more and more carbon dioxide dissolving in it; some we simply can't anticipate, and we have no way of knowing how serious those might be.

But even given these serious drawbacks, more and more scientists are listening seriously to techno-proposals, and agreeing that, at the very least, they should be researched. Some of the most talked-about schemes are profiled over the next few pages. While these ideas are all relatively recent, the notion of tampering with the climate is by no means a new one. In 1970 the government of the Soviet Union, worried about the drying up of the Aral Sea by irrigation, proposed reversing a substantial amount of the flow of the Ob, Yenisei, and Pechora rivers to refill the lake. They were right to worry: today the Aral is a tiny fraction of what it once was.

But the plan was terribly flawed. Reversing the flow of those rivers to the Arctic would have made the ocean saltier, reduced the amount of ice, raised the amount of sunlight absorbed (because open sea water is less reflective than ice), and, as we know now, given global warming a kick-start.

And how about this excerpt from *Restoring the Quality of Our Environment*, a report to President Lyndon Johnson by the Environmental Pollution Panel of the President's Science Advisory Committee in 1965:

The possibilities of deliberately bringing about countervailing climatic changes need to be thoroughly explored. A change in the radiation balance in the opposite direction to that which might result from the increase of atmospheric CO_2 could be produced by raising the albedo, or reflectivity, of the earth. Such a change in albedo could be brought about, for example by spreading very small reflecting particles over large oceanic areas. The particles should be sufficiently buoyant so that they will remain close to the sea surface and they should have a high reflectivity, so that even partial covering of the surface would be adequate to produce a marked change in the amount of reflected sunlight. Rough estimates indicate that enough particles partially to cover a square mile could be produced for perhaps one hundred dollars.

Of course, times have changed since the 1960s and 1970s. Or have they?

We are warming the planet, we're warming it faster than ever before, and most important, we really don't know what the results will be.

GOING DOWN
THE TUBES

IN 2007, SIR JAMES LOVELOCK AND CHRIS RAPLEY, HEAD OF THE
Science Museum in London, announced a desperate scheme for removing carbon
dioxide from the atmosphere: a new technology that, although untried and risky,
the authors felt they had to advance because, in their opinion, things are just so
desperate. Their letter to the science journal *Nature* was titled "Helping the Earth
to Cure Itself via the Oceans."

Their idea is to lower groups of huge tubes, tethered together, into the ocean.
The vertical tubes, something like 100 or even 200 metres long and 10 metres in
diameter, would be equipped with a simple one-way valve near the bottom (like
the valves in your arms and legs that prevent blood from slipping back between
heartbeats). Every time the tubes bob up and down in the waves, water from the
depths will ride up inside the tube, higher and higher, eventually reaching the
surface. It's well known that deep water is much richer in nutrients than surface
water, and spilling those nutrients onto the "relatively barren waters at the surface"

▶ In James Lovelock and Chris Rapley's scheme (as envisioned here by Atmocean), deep sea
water would rise slowly through large tubes, bringing additional nutrients to the surface and
stimulating the growth of phytoplankton, which would then absorb carbon dioxide.

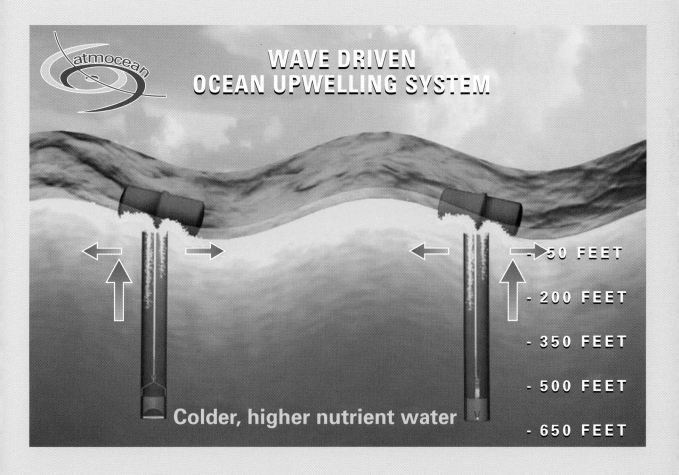

Early data suggest that the amount of carbon dioxide that actually descends deep enough varies dramatically from one part of the ocean to another.

(as Rapley and Lovelock put it) should encourage the growth of the single-celled plants called phytoplankton, which take up vast amounts of CO_2 as they bloom.

The issue then is to get that absorbed carbon dioxide out of there, back down to the depths, removing it from the atmosphere for a very long time. This should happen when the plankton die and their bodies fall slowly to the bottom, or when they're eaten by micropredators, whose feces then descend to the bottom.

Critics have questioned the potential of the idea from a number of points of view. Some wonder if carbon dioxide from deep waters might actually be brought back to the surface by the tubes, negating whatever benefit was derived from sinking the gas. Could the tubes stand the wear and tear? Stepping back, is it wise even to think about technologies that tamper with a natural system that is guaranteed to be more complicated than we know, and maybe even can imagine?

The idea of enhancing the activity of organisms at the surface so they take up way more CO_2 than they would have otherwise, then letting them sink to the bottom, carrying their carbon with them, is not new. Several groups of researchers have been interested in the idea of seeding the ocean's surface with a fine powder of iron, because there's good evidence that while nutrients aren't exactly rich at the surface, low levels of iron are growth limiting. A number of small-scale experiments have shown that dumping iron into the ocean can cause a bloom of phytoplankton. But once the phytoplankton have absorbed CO_2, does the carbon sink to

Atmocean will have to be cautious: companies have tried to use unproven technologies to claim carbon offsets before, but were forestalled by objections.

at least 1000 metres below the surface, the depth where we can be confident it isn't going to circulate back up? What other effects might a large-scale iron fertilization program have? At the moment, no one knows. Early data suggest that the amount of carbon dioxide that actually descends deep enough varies dramatically from one part of the ocean to another, for reasons that aren't yet clear. But you can bet the idea of encouraging phytoplankton growth is attracting a lot of attention.

Atmocean, a company based in New Mexico, has already leapt on the Lovelock/ Rapley idea and is publicizing its patent-pending technology to make the idea work. On its website, www.atmocean.com, the company predicts,

When fully deployed, our 3 m diameter by 200 m deep pumps spaced 2 km apart will be positioned across 80% of the world's oceans … Pumps are connected one-to-the-next at the base 200 meters deep, forming large arrays which maintain position from the sea-anchor effect … Arrays will be deployed outside of the 200 mile territorial limit to avoid busy shipping lanes.

Atmocean hopes to get into the carbon offset business (you pay the company to capture carbon dioxide on your behalf), but that is still years away, even if the technology works, which isn't at all certain right now. Atmocean will have to be cautious: companies have tried to use unproven technologies to claim carbon offsets before, but were forestalled by objections from scientists and environmentalists.

THE PROPHETS

Despairing for the Earth

Sir James Lovelock

He has been characterized as the Gandhi of science, lauded for his radical thinking by some, ridiculed by others. He forever changed the perception of the word *Gaia* (the Greek god of the earth) by using it to describe his concept that earth acts as if it were a living thing. According to his Gaia hypothesis, all forms of life on earth *plus* inanimate things like rocks, oceans, and the atmosphere work together to keep constant certain critical factors in the environment: temperature, oxidation state, acidity. Life on earth conspires to maintain conditions favourable to itself.

The idea that rocks, earth, and water participate in this live planet is a nonstarter for many, but Lovelock argues comparison to a giant redwood. Most of the tree, which no one would argue is not living, is dead tissue, the wood. If the tree can be 99 percent inanimate, why can't the earth?

I remember doing a couple of radio interviews with James Lovelock in the 1980s. It wasn't just his ideas that were intriguing, it was the fact that he was an "independent" scientist, not attached to any research institution. He has no time for the way science is done in academic labs, arguing that academe stifles scientific creativity. Lovelock doesn't worry about whom he might be offending or annoying with comments like that: environmentalists don't like it when he argues that nuclear power is really the only energy option; he contends that living through World War II was actually exhilarating; and now he says that it's too late—humanity is doomed.

Lovelock thinks that global warming has already gone too far, that temperatures are going to rise out of control; there will be mass migrations of tens of millions of people northward to countries where the temperatures are still bearable; the Sahara Desert will invade Europe; vast areas of the globe will be unsuitable for life, let alone cultivation; and the current world population of six-plus billion will be savagely reduced—by the new, overheated Gaia—to half a billion. He spells it out in his newest book, *The Revenge of Gaia.* Lovelock has thrown in the towel, and as a result, is dismissive of all kinds of activities that are promoted as saving the planet. Maybe, he suggests, some of this might have helped if it had been started in the sixties. But it's too late now.

Sir James somehow manages to keep a happy face, and he hasn't lost his innovative touch. Out of his despair for the earth came a suggestion in 2007 for a new technique to reduce carbon dioxide: tubes in the ocean.

If the tree
can be
99 percent
inanimate,
why can't
the earth?

Credit: Bruno Cornby—EFN—Environmentalists for Nuclear Energy

The Coccolithophores

Check out these two photographs: could they be more dissimilar? One is an image from a scanning electron microscope, the other a satellite photo. The scales they depict range from millions of metres to millionths of a metre, but the subject of both is the same: a sensational species called *Emiliania huxleyi*. It is a coccolithophore, one of the myriad species of free-floating microscopic plants in the ocean called phytoplankton. These coccolithophores are unique because of the tiny—but elegant—calcium carbonate plates they cover themselves with, the purpose of which is still not clear. (It is these plates that reflect enough sunlight to change the colour of the water and make the bloom visible.)

▼ Untold numbers of coccolithophores discolour the water off the south coast of Newfoundland (Nova Scotia and Prince Edward Island are in the background).

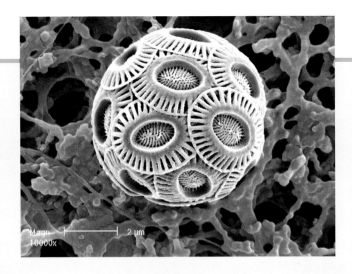

◄ A coccolithophore like this one may have thirty or more hubcap-like plates on its surface. Many more are shed into the water. When the organisms themselves number in the trillions, there can be hundreds of trillions of plates, enough to change the colour of the ocean.

In the right conditions, this species is able to reproduce at an unbelievable rate, creating vast blooms on the ocean surface, like the one off the coast of Newfoundland in the photo on the left. It's really impossible to judge the number of individual organisms there, but it would be trillions and trillions, at least.

These organisms are known for more than their occasional population boom—they also play a role in determining the earth's climate. For one thing, although their processing of carbon dioxide is complicated, the net effect of their population growth and death is to capture and drag deep into the ocean huge amounts of carbon, at least temporarily removing it from the atmosphere. Second, their plates are mirror-like, and blooms such as the one off Newfoundland reflect a significant amount of sunlight, reducing warming in at least that local area. But third and most important, organisms like *E. huxleyi* emit the gas dimethyl sulphide, which, once in the atmosphere, reacts to form other sulphur compounds, including sulphur dioxide. Sulphur dioxide, in turn, is a very effective nucleating agent, that is, water droplets will adhere to it, and clouds will form. Clouds insulate the earth but reflect sunlight as well, so these population explosions of *E. huxleyi* likely affect global warming in more than one way. But how large, or consistent, is their effect, and could it be prolonged or strengthened somehow? Those are good questions.

It's also true that we are just learning about these organisms. Fears had been widespread that as more CO_2 dissolved in the oceans and made them more acidic, *E. huxleyi* and related creatures would be less able to incorporate carbon dioxide into their shells, yet recent research seems to indicate that the opposite is true—that with more CO_2, they will add more calcium carbonate. And previous studies have suggested that these organisms have coped with changes in the carbon chemistry of the oceans in the past.

LOOKING AT CLOUDS FROM BOTH SIDES NOW

JOHN LATHAM, A SCIENTIST AT THE National Center for Atmospheric Research in Boulder, Colorado, tells the story of a time years ago when he and his son were looking at clouds, and his son asked him why they were light at the top and dark at the bottom. Latham replied, "Because they are mirrors for incoming sunlight."

Believe it or not, this brief exchange planted the idea in Latham's head that the mirror quality of clouds could be used in the fight against global warming. Specifically, thin, low-level clouds called marine stratocumulus, which are abundant over all the world's oceans but especially in the southern hemisphere, are efficient

▶ Thousands of these "cloud ships" would be needed to forestall uncontrollable rises in temperature.

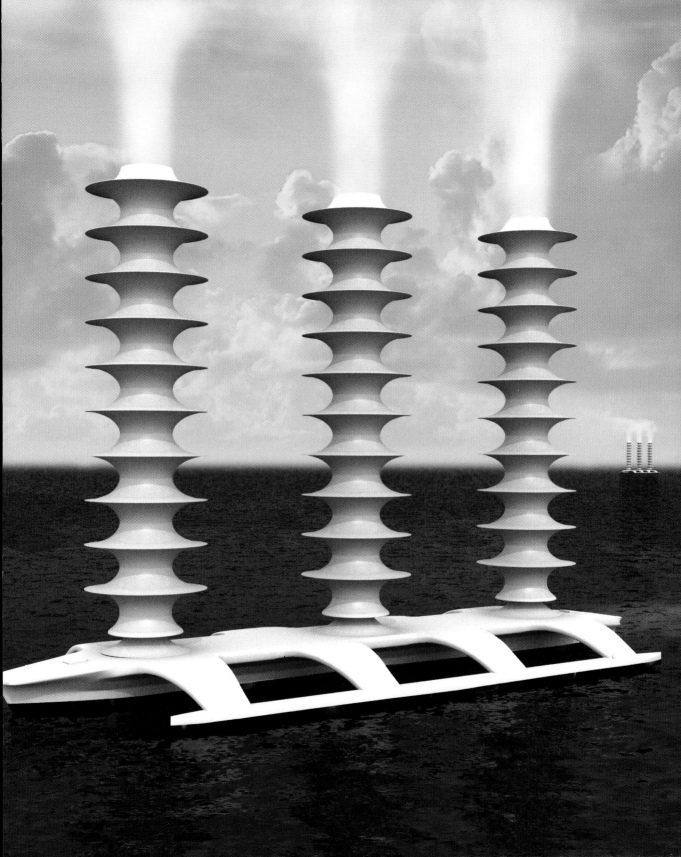

"It is urgent to start work on all these ideas as soon as possible to have them available if disaster is obviously looming."
STEPHEN SALTER

▲ In this future scenario, robot ships wander the southern seas, spraying ocean water into the air to build up cloud cover.

reflectors of incoming solar radiation. Latham speculated in a scientific article in 1990 that boosting the numbers of cloud droplets over the oceans (which are normally much fewer than over land) could neutralize the warming induced by increasing levels of carbon dioxide. In fact, if the reflectivity of those clouds could be increased by just 2 or 3 percent (from 50 to 52 percent), the reduction in sunlight reaching the earth would compensate for a doubling of carbon dioxide, an amount that has been seen as a critical turning point.

"We're not trying to create new clouds; we're not trying to extend the coverage of these clouds. We're simply trying to increase the number of droplets inside them. If we can make the clouds consist of, roughly speaking, a few big drops, instead of a lot of little drops, then they will, for good physical reasons, be better mirrors."

There is a happy coincidence here: the best clouds for the purposes of sunlight reflection are those over the oceans, and the oceans are full of salt particles, which, if you could just boost them into the atmosphere, would be perfect nucleation agents, that is, tiny particles that attract water molecules and encourage them to grow into the droplets that form clouds. As long as you can get droplets of ocean water into the air, they will evaporate.

So how do you do that? Stephen Salter at the University of Edinburgh has proposed using autonomous ocean-going vessels with huge funnel-like stacks on their decks. Inside the stacks, vertically oriented turbines would propel ocean spray into the air. To be most effective, they'd have to be able to do that at a rate of about 50 cubic metres of water per day. Or, to put it another way, they'd have to spray enough saltwater that a cupful of salt is launched for every square kilometre of ocean, every day.

The funnels would spin in the wind not only to propel the boats but also to generate the power to convert seawater to a fine mist (particles about a millionth of a metre in diameter), which, as it rises, evaporates, leaving the tiny salt crystals behind.

Even though this scheme has met with the usual cautions (or outright condemnations) about tampering with nature, it does have the virtue of merely imitating what already happens: ocean waves toss droplets into the air all the time. The mechanism would, of course, have to be designed so that the salt stays aloft long enough to help create clouds, but not so long as to reach landfall and turn farmland into salt flats.

No people would be on board—the ships would be satellite controlled—and when you consider the fact that it's estimated that thousands of them might be necessary to offset that dreaded two-degree rise in global temperature, the technical challenge is clear.

According to Stephen Salter, "We need to invent, design, and develop the spray mechanism that uses the least amount of energy to get the spray that we want." This project could easily cost tens of billions of dollars, but over decades, and if it worked, this would be money incredibly well spent. Tentative, small-scale tests still need to be done, studies to determine if indeed there might be environmental risks, but in this case, as with most of the geo-engineering ideas, it might be irresponsible not to at least consider it.

"The [U.K.] Met Office global climate model suggested very strongly that this technique would work and is a strong enough effect to produce a cooling that hopefully could hold the earth's temperature constant for a few decades while some new form—a clean form—of energy is found," Salter explains. "So I think it is urgent to start work on all these ideas as soon as possible. Not to deploy them until we're absolutely sure they would work and not do any damage, but to have them available if disaster is obviously looming."

THE HUMAN VOLCANO

WHEN MT. PINATUBO ERUPTED IN 1991, it spewed tonnes of gases and ash into the atmosphere. The blast was ten times the size of the eruption of Mt. St. Helens, and second only to the Katmai-Novarupta volcano eruption in Alaska in 1912. The result? The earth actually cooled, by about six-tenths of a degree Celsius, during 1992 and 1993. Scientists believe that cooling effect was largely due to the huge amounts of sulphur dioxide hurled into the upper atmosphere. Extremely tiny droplets of SO_2 scatter incoming sunlight, so that much of it is knocked sideways or backwards and never reaches the earth. If the SO_2 droplets are launched high enough, into the stratosphere, they can circulate around the globe for months. And if there are enough of them, as after Pinatubo, they will cool the earth.

Inspired by this impressive natural demonstration of climate control, scientists began to wonder a few years ago whether this effect was something that could be engineered. What if we injected vast amounts of sulphate aerosols into the upper atmosphere? Could we slow down global warming?

Ken Caldeira is a climate scientist at Stanford University's Carnegie Institute. He was skeptical when he first heard these ideas being floated, but

▶ Mt. Pinatubo's blast was ten times the size of the eruption of Mt. St. Helens and led to a slight cooling of the earth.

▲ Ken Caldeira's computer simulations surprised many by showing that the "human volcano" could actually work.

"The idea is if we turn up the heat with carbon dioxide, we can turn down the sunlight and bring the earth's temperature back to where it was before."

KEN CALDEIRA

decided to test them. "I actually started doing the simulations to show that these ideas were bad because I thought, well, that you're still going to have a big regional climate change, [but] the seasonal changes would be different. But as we did simulations, they seemed to show that it would actually work."

One of Caldeira's most amazing findings was that the necessary changes in solar radiation would be relatively easy to accomplish: "We're only talking about deflecting 1 to 2 percent of the sunlight. Say 2 percent. So you might not even notice it. It might make the clouds look a little hazier."

So sulphates in the upper atmosphere could slow or even stop global warming, at least temporarily. But significant technical issues need to be addressed, one of which is how to get the tiny aerosol particles just right. If they're the wrong size, or have the wrong electric charge on their surfaces, they'll clump together and be dragged back to earth by gravity. That's one problem. Another is, how would we get that sulphur up there, without having the explosive power of volcanoes at our disposal? Some have suggested rockets; Sir Richard Branson has toyed with the idea of burning fuel that is 1 percent richer in sulphur in jetliners, so we could accomplish what's needed with hardly a hesitation in our daily lives. One suggestion has even been to spray the particles from gigantic fire hoses suspended from balloons or dirigibles. Ken Caldeira isn't sure what the vehicle should be, but he

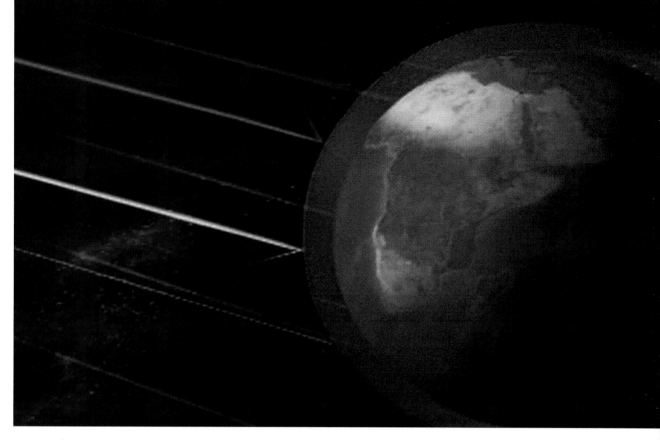

▲ The idea is to populate the upper atmosphere with molecules that will scatter and/or reflect sunlight away from the earth.

does know the size of the cargo: "Our calculations show that to solve the problem for the whole world, you would need about one tanker truck every ten minutes."

At first that sounds impossible. But Caldeira points out that power plants are already emitting twenty times that amount of sulphur, and muses that we could accomplish two valuable ends at the same time: by capturing emissions of sulphur compounds, which contribute to acid rain in the lower atmosphere, and injecting them into the upper atmosphere, we would both reduce emissions and mitigate global warming.

However, Caldeira and others who are thinking about geo-engineering are not unaware that any untried, large-scale technology comes with risks. Some computer simulations have shown, for instance, that sulphate aerosols in the stratosphere would, through a series of chemical steps, cause significant ozone depletion at both poles, exactly the thing we combated by banning chlorofluorocarbons in aerosol spray cans. The impact might even delay the recovery of the Antarctic ozone hole by up to seventy years. If nothing else, this unwanted side effect demonstrates the quagmire that geo-engineers have to avoid.

Salps

Polar bears and penguins may be the poster animals for global warming, but in the end, they won't be the most important organisms affected by climate change. Instead, animals that we tend to completely overlook will likely be the ones most intimately involved as global temperatures rise. The salp is a perfect example. Too small to have captured the public's imagination, and definitely not cute enough to cause anyone to go "Awwwwww," salps nonetheless could be the most important creatures on the planet as we struggle with an atmosphere brimming with carbon dioxide. At best, salps are about 10 centimetres long, but they make up for the lack of heft by their ability to form swarms numbering billions of individuals.

They are sea animals, and they will swarm in such numbers in response to blooms of phytoplankton, the microscopic plants that hover near the ocean's surface. Salps consume vast numbers of phytoplankton, but the strange thing is, nothing appears to eat salps. They have been described, memorably, as "the marine equivalent of inedible cows."

But wait—this could be a very good thing. Biologists at Woods Hole Oceanographic Institution, especially Laurence Madin, have discovered huge swarms of these creatures, and have concluded that they are capable of transporting vast amounts of carbon dioxide from the atmosphere to the deep ocean.

These swarms are unbelievable. One covered about 100,000 square kilometres of ocean, contained trillions of salps, and accomplished two amazing things: it consumed about three-quarters of all the phytoplankton on the surface (including all the carbon the plants absorbed from the atmosphere by photosynthesis) and shipped 4000 tonnes of carbon to the ocean's depths. Every day!

What sets salps apart from their micro-animal competitors is their feces. Fast-moving and heavy, they sink fast enough that the carbon they contain doesn't make it back to the surface. "Big packages that sink fast" is how Madin describes the salps' feces.

Salps hang around several hundred metres deep during the day, coming to the surface

These swarms are unbelievable. One covered about 100,000 square kilometres of ocean.

only to feed at night. They end up releasing their feces when they've descended again, and those feces sink at a rate of 1000 metres a day; even their bodies—also carbon-containing—descend more than 400 metres a day.

Who knew that the excretory habits of a small, unremarkable creature could loom large in the fight against global warming?

▼ A salp: beautiful in its own way.

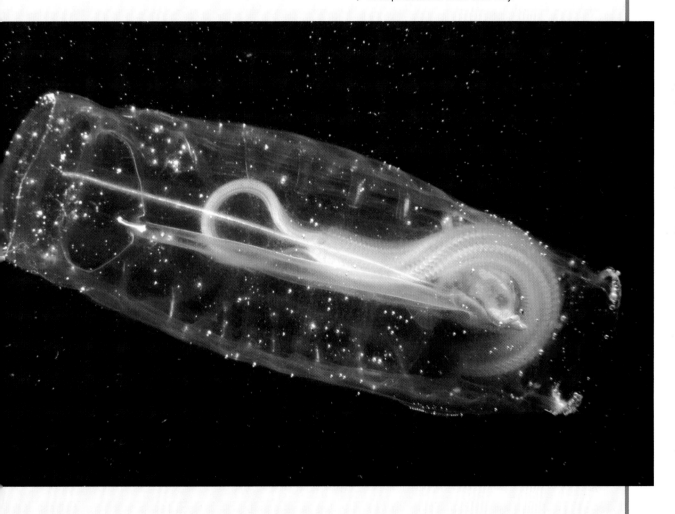

SPACE IS THE PLACE

INJECTING PARTICLES INTO THE UPPER ATMOSPHERE IS ONLY ONE possibility; engineers have also considered putting reflecting or scattering particles into low earth orbit. The advantage here is that the particles would likely stay in place longer than they would in the atmosphere. The problem is it would be much more expensive to get them there.

But as long as we're dreaming about the ideal technological solution—and I emphasize the word *dreaming*—Roger Angel of the University of Arizona has an idea. An astronomer by trade, Angel has earned the reputation of coming up with dramatic, even revolutionary ideas. The new Large Binocular Telescope is one of his most profound. But Roger Angel has also thought about ways to stop global warming in its tracks: "People have talked about things in earth orbit, and I'm looking at the most distant solution, which is this stable point in space which is a million miles toward the sun. If you put something in orbit there it will stay in line with the sun pretty much. So it's one place you can put a shade."

▶ Getting the shades (above) out to where they belong is the problem. Angel suspects that electromagnetic launching from deep in the earth would, in the end, be cheaper than launching by rocket (but still very expensive).

◀ Roger Angel is looking far beyond the earth's atmosphere to provide relief.

"To reduce the intensity of the sun, you need something between you and the sun to block some of that light."

ROGER ANGEL

Angel figures a 2 percent reduction in the sun's rays would cool the earth enough to bring us back to preindustrial times, before the era of carbon dioxide. His idea: launch sixteen trillion glass shades into deep space, arranged in a cloud 100,000 kilometres long and a few thousand kilometres wide. Each of the glass shades, although about half a metre across, would be incredibly thin, and would allow about 90 percent of the incident sunlight to pass through unimpeded: "It's almost impossible to imagine glass this thin, but it's so thin it would just blow away in the breeze. So the glass I'm looking at would be one-twentieth of one-thousandth of an inch thick. And it's actually, in miniature, a little spacecraft that has a little brain and some solar cells."

The glass shades, each no heavier than a butterfly, would need that brain and those solar cells to maintain their position in space and resist being displaced by the radiation pressure of sunlight. The technological challenge of manufacturing trillions of these shades is dwarfed by the task of getting them to where they're supposed to be. They collectively weigh close to twenty million tonnes, way too much to afford launches with traditional rockets. Instead, Angel envisions using electromagnetic space launchers, but even so the cost would be a staggering one trillion dollars. And the shades would have to be renewed every fifty years.

Some of the wilder schemes described earlier in this chapter—spraying sulphates into the stratosphere, scattering iron on the ocean, sinking tubes into the ocean—share one daunting characteristic with Angel's idea: they all threaten to bring about unwanted and unanticipated side effects. When you consider that such side effects could be global in scale, serious consideration and research are needed before any of these technologies are deployed.

On the other hand, more scientists now think that such dramatic schemes should at the very least be researched. Their reasoning is that not only have international plans to reduce carbon emissions failed, but emissions have risen as if no such plans or agreements had ever been made. These scientists are, if not already convinced that dramatic steps need to be taken, at least ready to contemplate such ideas.

But there is a middle ground: new technologies that are dramatic in their goals but benign in the sense that if they fail, they would cause no dramatic, negative, unforeseen impacts on the environment. Foremost among these technologies are ideas for removing carbon dioxide from the air. This idea has at least two variations, described over the next few pages. One is to grab the carbon dioxide as it's coming out of the smokestack; the other is to develop devices that will absorb the gas right from the air, anywhere on the planet.

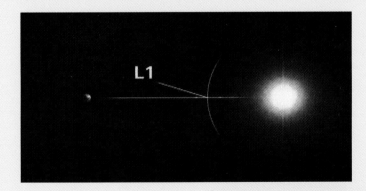

◀ L1, Lagrange point 1, is the place where Roger Angel would send his mini-shades.

▶ They would remain steady there, partly because the sun's and earth's gravity conspire to immobilize them, partly because each mini-shade has three solar panels and a small computer on board to be able to manoeuvre.

THE PROPHETS

How a Trip to Space
Changes Your View of Earth

Roberta Bondar

We are visual creatures. It has been estimated that fully 50 percent of our brains are devoted to processing the information flowing from our eyes. It makes sense then that those environmental changes we can see have the greatest impact on us: clear-cut forests, glaciers in retreat, forests devastated by the mountain pine beetle. It's a safe bet that in the future we will be witness to environmental change on an unfamiliar scale. Studies have suggested that as the world warms, Canada will be particularly hard hit. Habitats will move north at a rate that will be determined by how fast the temperature increases. Some computer models put that rate at a kilometre a year or even faster for at least a third of our terrestrial habitats. You might not notice much of a difference

▶ The Dalton Range in Kluane National Park, Yukon.

The Ward Hunt ice shelf in Quttinirpaaq National Park, Nunavut.

over a couple of years, but in a decade you certainly will. Species you were once familiar with will have moved on (or, if they can't migrate quickly enough, will simply be gone). Those that are more adaptable will move in to replace them.

In Europe there's already evidence of this mass migration happening: birds, butterflies, and alpine plants are shifting their range or, in the case of the birds, breeding earlier in the year. But remember this is still early, and we really cannot know how it's all going to turn out. What we do know is that there will be change. The natural world we see around us will look different.

We will have to adapt to those changes, but at the same time, we need a record of what we've had in the past and what we have now. The phrase "shifting baselines" refers to the risk that as things change, we simply accept them, forgetting what things were like before. To avoid falling into that shifting baselines trap, we need to record the natural world. One person who is doing that with great skill is Dr. Roberta Bondar.

Dr. Bondar is a remarkable person, someone who seems to have had more careers than possible to squeeze into one human lifetime. She, of course, was Canada's first woman astronaut; she is a medical doctor, a scientist, an author, a speaker, a university chancellor, and in some ways most importantly, a photographer. I say "importantly" not to downplay her other accomplishments, but because for us visual beings, anyone who has the ability to capture the fine-grained beauty of the natural world is someone whose work should—and does—attract our attention.

In her 2000 book *Passionate Vision: Discovering Canada's National Parks*, Dr. Bondar both writes about and portrays Canada before the most dramatic effects of climate change have been seen:

There are no people in my portraits, or things made by people. To see the Earth without humans can be disconcerting, but it draws attention to our presence. We can learn to see precious plants, trusting animals and delicate butterflies, which live from season to season, year to year, facing the hardships of weather, fighting to survive.

It is very real and clear from the space perspective that the forces of nature will keep reshaping the planet. Plants, animals and humans will constantly remodel and renew to meet the challenges or we will all vanish. But our environment cannot adapt quickly enough to compensate for our alien intrusion. Because humans have

"We must show our respect and admiration for our natural world and work for peace."
ROBERTA BONDAR

developed frightening technologies and have evolved quickly into a resource-depleting species, we, and we alone, have the ultimate responsibility to protect the Earth, and each other, from ourselves. We must show our respect and admiration for our natural world and work for peace, not destruction and extinction.

In my view, Canada's national parks system is a start in the right direction. There are at least four reasons to set aside and protect the land and the sea. First, protecting bio-diversity, a mix of plants, animals and micro-organisms, ensures that they can continue to evolve and maintain a healthy gene pool. Second, watching for shifts in the growth of other creatures can warn humans of existing and impending change. Third, learning how other life forms use sunlight, soil and water to respond to long-term weather changes can help humans to adapt, too. Unlike these life forms, which must seek a better place or die, we humans can cross over many habitats because of our technical ability. Last, protecting the parks enables the creatures that live and cycle through these areas to continue to do so unimpeded by humans.

Although we are an integral part of the environment, we are also observers and agents of change. We can induce and produce change in the environment, positively or negatively. Attempting to hold the environment in a steady state may withdraw the opportunity for natural evolution but we must try to protect other life forms from the forces of our technology, and the pressure of our sheer numbers.

Canada is unique in its large undeveloped and relatively untouched tracts of land and water. That is reason enough to be actively caring for our natural environment and heritage. The message is clear. If we do not protect the environment of our planet, we eventually will fail to keep our bodies, and our souls, nourished.

A GERM
OF AN IDEA

IN A LAB IN QUEBEC CITY, SCIENTISTS HAVE invented a bioreactor that is designed to attach to a factory smokestack. They are hoping it can capture up to 90 percent of the CO_2 that would otherwise escape into the air. The cool thing about it is that they have turned to biology to make it work.

When Dr. Sylvie Fradette was a student at Laval University, she led a team that isolated an enzyme from bacteria that was beautifully suited to industries trying to reduce their carbon emissions. Fradette explains, "Our technology is based on the use of an enzyme which everybody has in his blood. This enzyme is responsible for the transformation of CO_2 into bicarbonate ion."

Enzymes are molecules that speed up chemical reactions. Every living cell is full of them—life would not be possible without them. Fradette's company, CO_2 Solution, uses an enzyme produced by the common gut bacterium *E. coli*. But adapting the enzyme to an industrial process is a tricky operation. In the lab prototype, smokestack gas enters from the bottom and rises through the bioreactor, while water flows in from the top. The enzyme is highly efficient, but the issue is to expose as much of the gas as possible to the

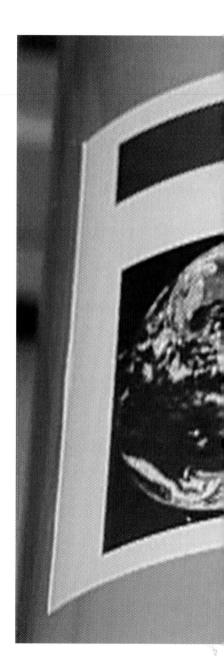

▲ Dr. Sylvie Fradette: biological tricks for a global problem.

"Our technology is based on the use of an enzyme which everybody has in his blood."
SYLVIE FRADETTE

enzyme molecules to be sure the maximum amount of CO_2 will be converted. The gas must be dissolved in water in order for the enzyme to react with it, and the design of the packing material to which the enzyme is attached is crucial—the carbon dioxide in the emissions must come in contact with the enzyme. There are always tradeoffs: it is a good thing to allow gas to flow through at a higher rate, but that might reduce the amount of contact between enzymes and gas. Ultimately, the final design has to maximize both flow rate and contact.

As with other CO_2-trapping technologies, capturing the gas from the air or smokestack emissions is only the first step. What happens to the CO_2 once it has been separated? In this case, the enzyme does double duty: it not only reacts with the carbon dioxide, it transforms it into calcium bicarbonate, which accumulates as a white powder. (Turning carbon dioxide into bicarbonate has been proposed before, but in a much more elaborate way, for instance by pumping the gas down into rocks containing mineral deposits rich in calcium. In fact, the white cliffs of Dover are beautiful examples of how this process was rampant millions of years ago.) This is one of the advantages of the CO_2 Solution process: a cement or pulp and paper plant could use the bicarbonate to generate the carbonate used in the industrial process, so there would be a closed circle: industry produces carbon dioxide, which is then transformed into bicarbonate and re-enters the manufacturing stream.

This is just one of many ways of trying to capture carbon dioxide at the stack, before it enters the atmosphere. But while that is feasible for large-scale industrial emissions, there are no practical technologies that could do the same for vehicle emissions, which make up somewhere between 15 and 20 percent of total global carbon dioxide emissions and are rising rapidly. People are thinking about this problem to be sure, but it's a tough one, at least partly because the standard internal combustion engine mixes fuel with air, diluting the carbon dioxide so much in the process that it's too difficult to capture it. Also, storing the carbon dioxide on board is a challenge because it is actually heavier than the fuel. However, researchers at Georgia Tech University have recently suggested that car engines could be built that would separate the hydrogen from fossil fuels and burn it, capturing and storing the carbon dioxide that remains. We're a long, long way from being able to do that, but researchers are out there working on the problem.

▲ In the end, the calcium bicarbonate that is left behind can be recycled through industrial processes.

▼ Perfecting the combination of bioreactor and packing material is a huge challenge.

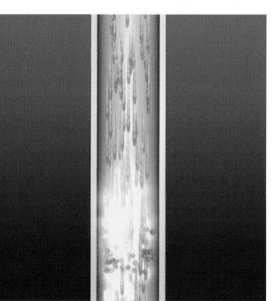

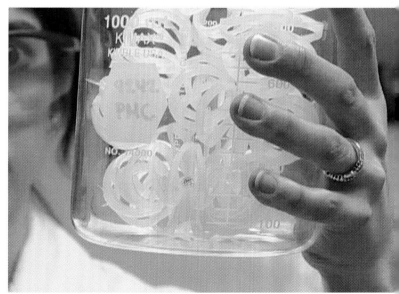

TREES ARE GOOD, BUT ARTIFICIAL TREES ARE BETTER

KLAUS LACKNER IS A PROFESSOR AT COLUMBIA UNIVERSITY IN NEW York. He, like a growing number of scientists, thinks that reducing our carbon dioxide emissions will not happen soon enough to slow and eventually halt the accumulation of CO_2 in the atmosphere before significant ecological damage occurs. So he is working on designs for devices that will actively remove carbon dioxide from the air: artificial trees.

We all know what leaves and needles on evergreens do: they absorb CO_2 from the air and use it to make sugars and starches by photosynthesis. This is what makes planting trees a useful strategy for reducing atmospheric carbon dioxide. But deciduous trees are virtually inactive in winter, and while reforestation is going on around the world, much of it is countered by deforestation, and atmospheric carbon dioxide is rising regardless.

With all that in mind, Klaus Lackner has been working to design an artificial tree, a device that would remove carbon dioxide from the atmosphere more aggressively than natural trees do. Believe it or not, an artificial tree has advantages, at least for the purposes of removing CO_2, over natural trees: "The similarity to what we are doing is that we have surfaces just like the leaves over which the air blows and the CO_2 is taken out. Our leaves can be stacked much more; they can get into each other's way with regard to the sun. That doesn't matter. So we can collect much more CO_2 at the sides of the tree like that than we would with a natural tree."

▶ An idealized portrait perhaps, but one possible realization of the artificial tree.

> "We have to make sure we have a carbon-neutral world economy. Therefore, for every tonne of carbon dioxide that is emitted, a tonne of carbon dioxide is taken care of and put away again."
>
> KLAUS LACKNER

▲ Surrounded by leafless winter trees, Klaus Lackner is thinking about powerful, year-round, night-and-day *synthetic* versions.

The idea may be barely off the drawing board, but the necessary elements are well understood: "As the air moves over the filter material—the 'leaves'—the CO_2 is grabbed and turned into sodium carbonate, and then the back step of the operation is to take the sodium carbonate, break it apart into sodium hydroxide and CO_2, and you get your CO_2 back. The trick is to establish the uptake rate which, with the overall design, ultimately tells you whether you want an air path which is very long or whether you get away with an airpath which is very short." There are other technical issues too: one is how to remove the carbon dioxide that quickly coats whatever material is used to absorb it.

The ultimate goal is to create an artificial tree about the size of a football goalpost, with several thousand leaves, that would have the incredible ability to take up an amount of carbon dioxide equivalent to the tailpipe emissions of four thousand cars. And these would be cars anywhere in the world, because CO_2 mixes evenly throughout the atmosphere.

If such a tree can be built, Lackner estimates that 250,000 of them would be needed to reach that idealized state of a carbon-neutral world. The cost, at least at the moment, is unknown, but neither Lackner nor others who are building their own carbon-removal systems think that the costs will be prohibitive.

▲ Plenty of lab work remains to be done to solidify the concept and prove that it's economically viable.

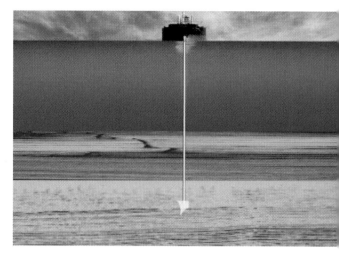

▲ Once trapped, carbon dioxide must be permanently removed from the atmosphere— perhaps below the seafloor.

◄ These plastic rings, coated with carbon dioxide–absorbing material like sodium hydroxide and packed together, are the guts of the artificial tree.

However, it's not just working out the technology that's a challenge. There's also the issue of dealing with the carbon dioxide once it's been trapped from the air. It could be buried underground: that's the way it's being done in Saskatchewan already, some thirty million tonnes annually. It could be deposited in the ocean—that's one of Lackner's thoughts: "Not into the ocean, below the ocean floor where the CO_2 is now denser than water and has no propensity to come back, and then ultimately you combine it chemically and make magnesium carbonate out of it."

An even wilder idea is to combine the captured carbon dioxide with hydrogen (perhaps generated by using electricity to split water molecules), to re-form it into a hydrocarbon. It is conceivable that such materials could be used as fuel. Burning them would generate carbon dioxide again, but this time the gas would not represent a net increase, because it had already been in the atmosphere once. I guess I don't need to say that we're a long way from reaching this point, the hitch being that the process requires so much energy that it's simply not worth doing—at least so far.

THREE
Green Projects

Some technologies to address the problem
of climate change are already up and running:
geothermal power, wind turbines, and hybrid cars.
At the other end of the spectrum, on a much
grander scale, are schemes for tinkering with
the climate directly. But there is plenty
of room between these extremes to experiment
and make existing technologies better.
Tweaking everything we already know to be
just a little more efficient, a little more adaptable,
might be the most important thing we can do.

TOWER OF POWER

IT'S EASY TO SEE WHY SOLAR POWER SEEMS like a sensible approach to sustainable energy generation: it's free, it's abundant (the amount of sunlight striking the earth in forty minutes is as much energy as the entire globe uses in a year!), and there are well-established ways of harnessing it. It does have drawbacks—they are called "night" and "cloudy days"—but even so, solar has not been used to anywhere near its potential. It's fair to say that solar is still in the experimental phase, and in a way that's a good thing: it encourages open-mindedness.

For instance, to most people solar power means photovoltaic cells, those devices that use clever electronics to convert the energy of photons from the sun to electricity. Although at the moment their use is mostly small scale—in homes to augment power from the electrical grid, and in specialized applications like roadside telephones—photovoltaics have huge potential to supply significant amounts of power in many countries around the world.

▶ At Sandia National Labs, they take their solar power seriously.

The engine is like the one in your car, but it burns sunlight instead of gasoline.

But photovoltaics are not the only route to solar power. At Sandia National Laboratories in Albuquerque, New Mexico, researchers have created two impressive demonstrations of how to take the "power" in solar power to a whole new level.

One is a technology that Sandia is developing, together with California's Stirling Energy Systems, to concentrate sunlight and use it to drive an engine that will generate electricity. There is an ancient legend that the great mathematician Archimedes foiled a Roman naval invasion of the port of Syracuse by focusing the sun's rays with soldiers' shields and setting the invading ships on fire. The story is probably not true, but the principle works. At Sandia, those giant dishes are not tracking satellites—they are tracking the sun. They follow the sun through the day, and each is covered with an array of curved mirrors that focus the sun's rays on a relatively uncomplicated four-cylinder engine. The engine contains hydrogen gas, which, when heated, drives the pistons in the engine and generates electricity. In a sense, the engine is like the one in

your car, but it burns sunlight instead of gasoline. And emits nothing. Each unit is relatively small, generating only about 25 kilowatts of energy, but a single acre of otherwise unused land can support eight such systems and power somewhere between ninety and one hundred typical homes.

Even better, this system is environmentally benign: the only water it uses is for washing the mirrors; each system takes up about the same space as a tree; and the major environmental impact is to create shade.

In February 2008, a system like this achieved a world record in energy conversion, hitting 31 percent net efficiency in changing sunlight to electricity, nearly 2 percent higher than the previous record. (Efficiency is measured by dividing the

▼ A sense of scale.

▲ The concentration of light in the power tower is enough to melt a hole in a brick!

electrical power generated by the total amount of solar energy hitting the dish.) Admittedly, it was a perfect day: temperatures near freezing and a daytime brightness about 8 percent higher than normal meant both the system's efficiency and the total available energy were at optimal levels. Even so, it was a significant step forward in an industry where every percentage point of efficiency is crucial.

Sandia is also developing a much more dramatic technology using something called the power tower. In this case, a flock of sun-tracking mirrors focuses on a single receiver at the top of a tower. There, the solar energy heats either steam or salt, and again that heat generates electricity. The salt is molten and can be stored for long periods after solar heating, allowing the continued production

of electricity on cloudy days or at night. The biggest difference between this and the individual mirror and engine technology is the sheer size and intensity: a single tower can generate 100 megawatts of electricity, enough to power fifty thousand homes. Of course, this system needs much more land, about 1000 acres. In the southwestern United States, where sufficient sunshine is guaranteed, there is plenty of land that could be set aside for power towers. One is already up and running near Seville, Spain.

For their next act, Sandia scientists are working on a device that would use solar power to split molecules of carbon dioxide into carbon monoxide and water; the monoxide would then become a building block to make new hydrocarbons. They call it the "Sunshine to Petrol" project.

Bugged by Global Warming

Most of the living things featured in this book are potential victims of global warming, unable to cope with the ecological changes wrought by higher temperatures and more acidic oceans. But in a perverse twist, the insect featured here has both benefited from global warming and is contributing *to* it. It is the mountain pine beetle, and people who live in British Columbia and Alberta are certainly familiar with it and the destruction it has caused.

The beetle is less than a centimetre long, about the size of a grain of rice. It lives under the bark of pine trees, particularly lodgepole pine, and most of the time is simply a normal part of forest ecology, preying on diseased or old trees. Masses of female beetles invade trees in the fall, carrying with them a fungus that inhibits one of the tree's only defences, the production of sticky resin. It usually doesn't take long for the tree to be overwhelmed and starve to death because the beetles, together with the

fungus, disrupt the tissues that bring food and water up the trunk to the limbs of the tree. Eventually, the trees' needles turn a tell-tale red. Red for dead.

The beetle has been around for thousands of years, but recent events have conspired to turn it from an incidental killer of occasional trees into a scourge in British Columbia. Winter weather has not been cold enough to kill off overwintering larvae, the summers have been hotter and drier, and the forests have contained unusually high numbers of mature pines. Pine beetle epidemics can be cut short if winter temperatures hover around minus 35 or minus 40 degrees Celsius for several consecutive days, but in recent years there have been very few sets of days like that. So the slowly warming world has favoured the survival of a larger number of beetles than before. The result has been an unprecedented destruction of pine trees. Millions have died, and it is predicted that 80 percent of the mature pines in British Columbia will be dead by the year 2013.

To make matters worse, Werner Kurz and his colleagues at the Canadian Forest Service have discovered that the beetles' impact on the forests could actually worsen

▶ The culprit (above left) and the damage done. Every red tree is a dead tree, and the beetles have already moved on to others.

global warming. If the rate of forest destruction continues, as it appears it will, then a vast amount of timber will stop growing, die, and start to decompose. When this happens, the trees stop taking up carbon dioxide and start releasing it back into the atmosphere. Of course, this is a natural, contemporary recycling process and so is quite different from venting carbon dioxide that has been trapped in fossil fuels for millions of years. But that doesn't mean we can ignore it. The problem is that in trying to estimate the impact of ever-increasing amounts of carbon dioxide into the atmosphere, scientists factor in how much of that CO_2 will be taken up by trees. In fact, tree planting is one way to mitigate the effects of too much CO_2.

Forests are usually considered to be carbon sinks, that is, they absorb more CO_2 than they produce. But when Kurz and his fellow scientists ran computer projections of the fate of the B.C. forests from 2000 to 2020, they found that those forests would act as carbon sinks only for the first two years, and after that would contribute to the atmospheric load of carbon dioxide. Why? They took into account the likely impact of forest fires (which release carbon dioxide from the trees they burn) and the fate of trees killed by pine beetles, whether they were harvested or allowed to decompose.

Even though all three processes release carbon rather than taking it up, the impact of decomposing trees killed by the beetles far exceeded that of forest fires. In fact, over the projected twenty-year period, the carbon emissions from beetle-damaged trees amounted to the equivalent of five years' worth of emissions from all transportation in Canada—a billion tonnes of carbon dioxide!

Many want to harvest as much of the wood as possible before it rots, although clear-cutting infested forests may simply set up the possibility of future infestations, if those clear-cuts end up being vast uniform forests of single-species trees all the same age. But if the wood is harvested, some advocate turning it into biofuel.

Jack Saddler, dean of the faculty of forestry at the University of British Columbia, argues that a billion cubic metres of lodgepole pine could be harvested by 2013, an amount never seen before. Why not turn it into biofuel? Without saying we should turn it into ethanol, he did point out that if only 25 percent of beetle-killed wood were ethanolized, it could supply some years' worth (anywhere between five and ten) of British Columbia's gasoline needs. There is a precedent: Canada exports more than a million tonnes of wood pellets to Europe every year. Made from stuff like sawdust, they're supposed to be clean-burning and efficient. And they don't compete with the human food supply like crop-based biofuels do. Fortunately, there's plenty of easily accessible sawdust, more than enough to make this plan work economically. But when your feedstock is beetle-infested pine trees standing in the remote woods, it's not so clear how that would work.

Where next for the beetles? Western Alberta is already feeling the heat, and the absolute worst-case scenario would be spread to the northern boreal forest. Most attempts in the past to stop the mountain pine beetle in its tracks have been unsuccessful. In the end, they might just run out of pine trees, at least in British Columbia, but by then, from the point of view of global warming, the damage will be done.

▼ Forests devastated by the mountain pine beetle, an insect that is both benefitting from and adding to global warming.

The slowly warming world has favoured the survival of a larger number of beetles than before.

LOOK UP,
LOOK WAAAAAY UP

WIND POWER IS A SIGNIFICANT PART OF THE PLAN TO DECREASE our reliance on fossil fuels. Wind farms are now common; the turbines themselves are being refined all the time (see the humpback turbine on page 122), but I bet you've never seen a wind turbine like this. It's called the Magenn Air Rotor System, or MARS. It's a neat invention, and sounds like it could do the job, but it is still an untried technology. Untried, but pretty cool.

MARS is a combination of a wind turbine and a helium balloon. It's tethered to the ground, and reeled out to about 300 metres altitude. The winds up there are well suited to the generation of electricity: they blow straight and steady, unimpeded by ground effects like hills or buildings. The blades along the cylinder turn in the wind, providing not just electricity but a major advantage aerodynamically. The cylinder spins like a golf ball, and this spinning, called the Magnus effect, gives the cylinder (and the golf ball) lift. (The dimples on the golf ball enhance this effect by creating turbulence in the layer of air that hugs the ball. The end result is that the ball goes farther, as long as it has backspin, not topspin.)

This lift from spinning stabilizes MARS when it's aloft; it's never supposed to tilt more than 45 degrees, no matter how strong the wind (hurricanes not included). The electricity generated by its spinning blades is transmitted by wire back down the tether, either directly to the grid or to be stored in batteries.

▶ It's a balloon! It's a turbine! It's a giant golf ball! Actually it's a very cool new technology.

Could these spinning, airborne turbines be a technology that will help wind energy take off?

WIND

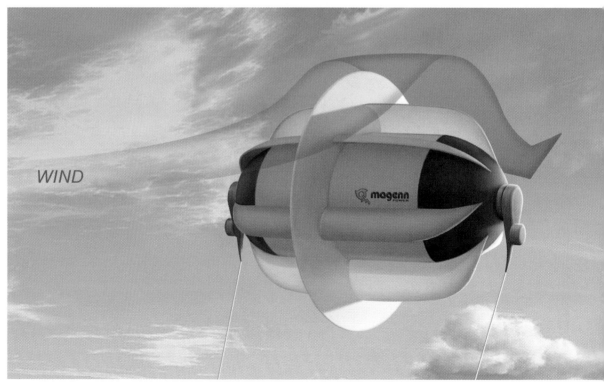

The ill-fated
Magnus airship
was a forerunner
of the MARS
turbines.
Cool technology,
but no great
need for it.
However, there
is a need
for wind
turbines
such as
MARS.

They aren't as efficient as the standard wind turbine, but MARS turbines are much more flexible: they can easily be hauled in, deflated, packed, and taken to a new site and launched again. That would make them ideal for use in disaster relief. They could be made in different sizes as well, even one that almost fits in a backpack, and would cost just a few hundred dollars to generate a few hundred watts. They could be used on islands or in remote communities, and even be packed together to create aerial wind farms. If the small-scale version is built, you could go camping with one! They're also touted to be friendlier to birds and bats than regular turbines because they turn much more slowly. That's what researchers at Magenn Power are projecting, anyway—they haven't made them yet.

A curious sidenote: Fred Ferguson is the inventor of the concept. He was briefly famous in the 1980s for his invention of the Magnus Airship, a blimp that spun on its axis to get that same Magnus lift effect. It was going to play the role of a heavy-duty helicopter, designed to lift close to 60 tonnes and travel at 90-plus kilometres an hour. The funding for the Magnus Airship came originally from the Star Wars program to develop space-based weapons in the United States. It was grounded when Star Wars funding ended, but Ferguson adapted the idea to MARS. There is no doubt he is inventive, and if you want to believe in genetic determinism, he is descended from Samuel Morse of telegraph fame.

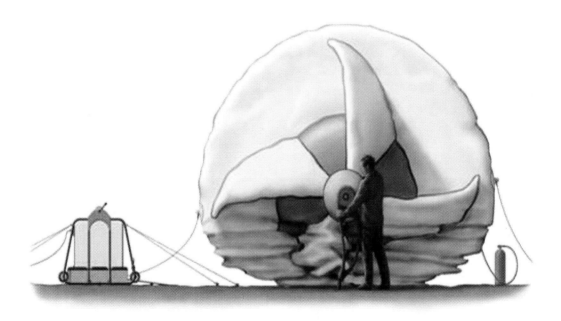

MAGENN AIR ROTOR SYSTEM

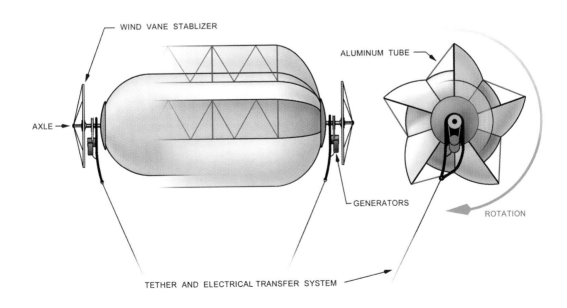

WIND VANE STABLIZER

ALUMINUM TUBE

AXLE

GENERATORS

ROTATION

TETHER AND ELECTRICAL TRANSFER SYSTEM

THE KITE SHIP

SOME INVENTIONS FOR SAVING THE PLANET are grandiose—you'll see some of those in this book. Others are incremental, relatively minor adjustments to existing technologies, but we're too early in the game to know whether one approach bests the other. In the end, we may need everything and anything we can get our hands on.

Take shipping. It transports 90 percent of trade goods worldwide. While the emissions from aviation—which make up about 2 percent of the world's total greenhouse gas emissions—have been targeted by environmental groups and critics like George Monbiot (see page 112), shipping has somehow stayed under the radar (it wasn't even included in the Kyoto Accord).

That little honeymoon ended when recent analyses of fuel consumption and engine size of the world's shipping fleet revealed that it was emitting more than twice as much CO_2 as airplanes, and was likely to increase that amount by something like 75 percent by 2020. Even though ships have become more fuel efficient over the last twenty years, it's clear they are going to be required to do much more.

▶ Its Skysail fully deployed, this ship is taking advantage of ocean winds to reduce fuel consumption.

▲ Kite ships are a novelty. Exactly how much impact they will have on the global fleet and its carbon emissions remains to be seen.

Enter MV *Beluga Skysails*. It is the world's first commercial ship to conserve fuel by flying a giant kite when it's on the high seas. It completed a long maiden voyage in early 2008, and the kite it deployed allowed it to rack up fuel savings of 10 to 15 percent. The ship travelled from Germany to Venezuela, then to the United States and back across the Atlantic to Norway, a grand total of nearly 12,000 nautical miles.

The kite has an area of 160 square metres, and it can fly up to 300 metres above the surface to catch more powerful winds (see MARS on page 102). The company believes the kite can generate five times as much power as a traditional sail. This inaugural voyage allowed the company to tweak the computer control of the kite. Most of the time the kite isn't allowed to fly free; instead, it carries out a prepro-grammed set of movements. Stefan Wrage, the managing director of Skysails, thinks this is the crucial secret: "We are manoeu-vring the kite, we are flying in patterns all the time and because we do this, we can get back much more thrust than we can with the normal sail. That's the true secret. It works all the time. Doesn't stay there static and stable."

It all comes down to physics. If the kite is flown in a figure-eight pattern, the speed of

the wind over the surface of the kite increases, and that can translate to huge increases in power. For instance, if the kite is experiencing a wind of about 12 metres per second, and by swooping and diving that speed can be increased to 20 metres, the kite will generate nearly three times the power.

There's more physics: the kite is not going to force the ship to keel over in the wind like the sails on an old-fashioned sailing vessel would, because the point of attachment is at deck level at the bow, rather than on a towering mast.

The next step is to deploy a 600-square-metre kite, which should deliver twice as much power and save twice as much fuel as the 160-square-metre kite. Even so, the biggest oil tankers in the world can be thirty, forty, or even fifty times bigger than the *Beluga Skysails*—what impact, if any, even *very large* kites would have on such behemoths remains to be seen. Nonetheless, Skysails is hoping to outfit some thirty ships in the next year or so.

Recent analyses of fuel consumption and engine size of the world's shipping fleet revealed that it was emitting more than twice as much CO_2 as airplanes.

"It is primarily a moral issue because people in the rich nations, which by and large are in the temperate parts of the world, are not those who will be hit hardest. And yet we are the ones who are most responsible for causing the problem."

GEORGE MONBIOT

The Truth Too Inconvenient Even for Al Gore

George Monbiot

George Monbiot makes people nervous. Even those who would call themselves environmentalists. He has this effect because he argues for greenhouse gas reductions that seem bizarrely exaggerated, then proceeds to build, step-by-step, the arguments for those reductions. When he's finished, you feel conflicted: you can't really make yourself believe what he's saying, but you can't find fault with it either.

Monbiot is an investigative journalist and author in England. He has devoted many of his columns in *The Guardian* newspaper to climate change, but he has also written a book called *Heat: How to Stop the Planet from Burning*. It's not a book about the steps you and I might take to reduce our personal carbon footprint. No, this is a book that declares that Canada, for instance, must reduce its CO_2 emissions by 94 percent from current levels. That is not a misprint: 94 percent is what Monbiot calculates we need to achieve. We're not going to get there by switching to compact fluorescents!

Monbiot is definitely at the hard-nosed end of the global warming spectrum. However, unlike James Lovelock, he hasn't completely abandoned hope. He's come up with a prescription for getting a grip on global warming; it's just that his prescription is bitter, bitter medicine indeed. And if you think he is holier-than-thou, a few sentences from his book should dispel that suspicion:

Most environmentalists—and I include myself in this—are hypocrites … By and large, whatever our beliefs may be, we consume as much as our incomes allow. Environmentalism is for other people. What this means is that changes of the kind I advocate … cannot take place without constraints which apply to everyone, rather than to everyone else.

Monbiot argues that carbon rationing is the only fair way to ensure that we cut emissions globally by a sufficient amount to avoid catastrophe. In that scheme, every country has its carbon allotment (based on its population), and it's up to each government to decide how that carbon should be distributed. Of course, *meeting* the carbon allotment is the key issue, and while Canada's 94 percent reduction may seem insane, reductions of that order would be necessary for most developed countries. Monbiot uses his book to lay out a strategy whereby these dramatic global cuts in carbon emissions could be made, "though by the skin of my teeth."

Here is part of a *Daily Planet* interview with George Monbiot:

George: It is remarkable the speed with which awareness of climate change has been rising, and people's preparedness to take action. It's happened within just the past year or two. Now at least we're seeing the Canadian government beginning to pay some sort of cognizance towards the problem. We're seeing stirrings in Congress in the U.S. We're seeing the European Union now set a new target for itself. We've even heard the Chinese government making some very positive sounds about the need to take action on carbon emissions. Things are happening very fast. Unfortunately, not fast enough.

Jay: So how fast do you think we have to move?

George: I believe that to have a high chance of preventing 2 degrees of global warming—and that's a critical figure because if it goes beyond that we end up really with runaway climate change that we can't do very much about—we require a 90 percent cut in the rich nations in carbon emissions between now and 2030. Ninety percent because globally we need to see a 60 percent cut. But if it's to be evenly distributed around the world, in other words if everyone on earth is to have the same entitlement to produce carbon dioxide and other greenhouse gases, then we have to take the bigger

THE PROPHETS
(continued)

share of the cut because at the moment we are producing more than the people in the less-rich nations.

Jay: Given that we in North America are energy hogs, that would suggest that we're going to have to suffer some serious lifestyle changes.

George: It does. And I think the truth too inconvenient even for Al Gore to articulate is that the American way of life is not sustainable, that there are some aspects which will really have to change if you're going to take this issue seriously. One of them is the size of the cars you drive. On the whole, they are much, much bigger than in most parts of the world. And much less fuel efficient. Another is the number of flights you take and aviation is one of the big, big climate change problems because there are no good substitutes for the way that planes fly currently on the horizon. There is no way of markedly improving their fuel efficiency.

Jay: George, you say it's the moral issue of the twenty-first century—why?

George: Because this issue, above all others, has the capacity to create a staggering amount of human suffering. We have seen predictions now coming from the top levels of climate science showing that there's a

huge potential for an expansion of drought zones in Africa and other parts of the tropics such that many hundreds of millions of people might find themselves unable to grow food any more, unable to support

themselves any more. We have the capacity for sea-level rises affecting hundreds of millions of people more. We could see a great deal of deaths caused by climate change. It is primarily a moral issue because people in the rich nations, which by and large are in the temperate parts of the world, are not those who will be hit hardest. And yet we are the ones who are most responsible for causing the problem.

BLEND ME A WING

ALTHOUGH PASSENGER aircraft are thought to represent only 2 percent of all greenhouse gas emissions, aviation is one of the fastest growing industries, and aircraft emissions are projected to double by 2030. The Intergovernmental Panel on Climate Change estimated that by

▲ Radical designs require radical testing.

2050 aviation could account for up to 10 percent of global carbon dioxide emissions. In other words, as countries and individuals struggle to reduce their carbon footprint, passenger airlines almost inevitably will increase theirs.

At the moment airlines are desperately trying to conserve fuel, as much to keep costs down as to protect the planet, but these measures, such as washing jet engines to keep them cutting smoothly through the air, or coasting into airports, are relatively trivial compared to the scale of the problem.

So what to do? The draconian option, championed by George Monbiot, is to stop flying. Thinking about going to Tuscany for your summer holidays? There's a nice cruise ship leaving Halifax

◄ Is this the future of flight?

▲ Traditional airplanes are two wings stuck on a tube. In this blended wing model, wings and fuselage are one, and the engines are mounted at the back and on top.

in July. And a good highway to Halifax from wherever you live. Or maybe you should just get some takeout pizza and a bottle of Chianti, and pretend.

At the other end of the spectrum is Sir Richard Branson, who has flown one of his Virgin Airlines 747s with one fuel tank containing biofuel (a mixture of oils from ba-bassu nuts and coconuts). He thinks biofuels, particularly those derived from algae, represent the future of the airline industry.

But there is a third approach that leaps beyond the idea of biofuels to reconsidering the shape of the airplane itself. Called the "blended wing," this plane is touted by some proponents as the hybrid car of aircraft. Certainly, something along the hybrid lines is needed, when you consider that in crossing the Atlantic, the average passenger jet consumes more fuel than a car would in fifty years. The blended wing turns the old idea of what an aircraft looks like upside down. Actually, it looks more like it's been melted down.

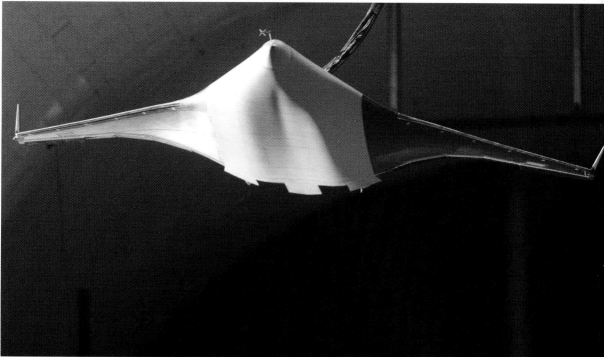

▲ The good news is that the blended wing models seem to perform well in wind tunnel tests. The bad news is that they're still being tested in wind tunnels.

▲ Professor Ann Dowling of the Silent Aircraft Initiative.

The old idea is called "tube-and-wing." The two wings are simply attached to the tubular fuselage. This design has worked well, but it has aerodynamic shortcomings. At least two research groups have been looking at a fundamental redesign of the tube-and-wing. Boeing is one, with its blended wing body, and a consortium of Cambridge University, MIT, and aerospace companies called the Silent Aircraft Initiative is the other. Its plane is called SAX-40, "Silent Aircraft eXperimental."

Silent? Professor Ann Dowling is at Cambridge University: "We've set ourselves a very strong challenge and that is if you start with a blank piece of paper, can one design an aircraft so people in the city should just not hear this aircraft as it flies overhead? So that was question one. Question two was, if you could design such an aircraft, what would be the impact on fuel burn?"

> ## "We were pleasantly surprised at just how easy this vehicle is to fly."
> ### DAN VICROY

It turns out that cutting down on aircraft noise brings with it better fuel efficiency, partly because the engines are placed on top of the plane to prevent sound from reaching the ground.

"We ingest the slow-moving air near the airframe into the engines which are placed here at the back. And that reduces the wake from behind the aircraft and gives us an improvement in fuel burn. As well, in a conventional tube-and-wing, the lift is generated only on the wings, and not on the cylinder, the fuselage, whereas in a blended wing body concept, all of the airframe contributes to lift. So we have a combination of lift generated on the centre body and very efficient outer wings."

There is convergent thinking happening here. Boeing is testing a model of its own blended wing aircraft at NASA's Langley Research Center. Because the design is so radically different from aircraft that have flown before, the testing has to start very near the beginning. So near the researchers weren't even sure how the scale model would fly. Of one test in which gas was pumped into the model to simulate thrust from its (nonexistent) engines, Dan Vicroy, a senior research engineer at NASA, said, "It's flying very well, actually. We were pleasantly surprised at just how easy this vehicle is to fly. We know how to predict how conventional tube-and-wing airplanes are going to fly, but with this type of design, we really have very limited experience."

A cautionary note here: we probably could make good use of fuel-efficient airlines right now, but we're not likely to see these new concepts for decades; funding is the major stumbling block. However, in the time it took to write about the blended wing, the pressure exerted on airlines (and their customers) by the rising price of oil has made even far-out concepts much more attractive.

A WHALE OF A TURBINE

OF THE MOST PROMISING REPLACE-ments for fossil fuels, wind and solar head the list, and of those, wind appears to be the more mature technology. And there is absolutely no doubt that it is plentiful. In fact, the British government has decided to go ahead with a plan to build a giant offshore wind turbine array, including seven thousand turbines over the next ten years, generating a quarter of the U.K.'s electricity needs. But is wind power genera-tion as efficient as it could possibly be? Likely not. And a research team in Canada is betting that lessons from nature—in partic-ular, the humpback whale—will help improve wind power. At first this looks like an unusual connection, but you always have to look beneath the surface.

Humpback whales are incredibly agile animals for their size, sometimes making abrupt turns as they pursue prey as small as tiny crustaceans. Scientists were pretty sure their extraordinarily long fins had something to do with their manoeuvrability, but they weren't sure exactly what. But then biologist Frank Fish (which would be a funny name except that whales are mammals) and his colleagues turned their attention to the bumps, called tubercles, on the leading edge of the humpback fin. At first this didn't make sense: putting bumps on the leading edge is a little like putting a series of hood ornaments

▶ Humpback whales, despite their bulk, are amazingly agile in the water. Part of the reason lies in the bumps on their fins.

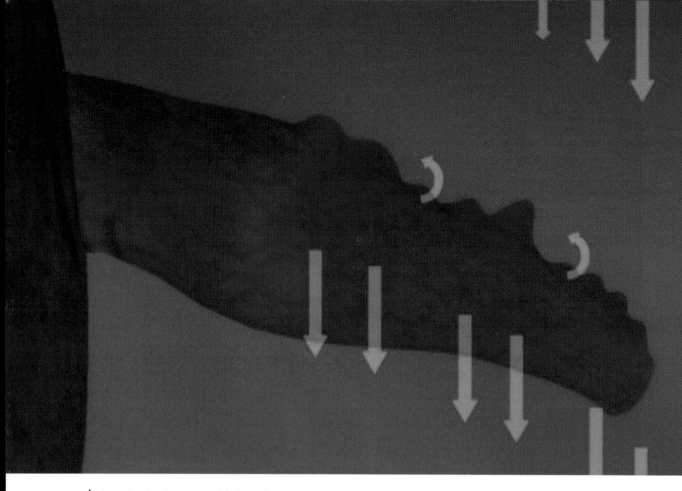

▲ Like the dimples on a golf ball moving through the air, the tubercles on the humpback fin give the fin a better grip on the water flowing over it.

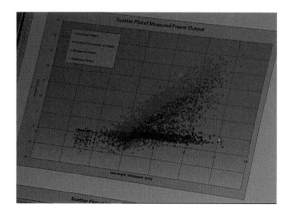

▲ Tests show that blades with tubercles capture more energy at moderate wind speeds.

on a car—you'd think they would just cause drag and slow the car down. But in fact the opposite was true.

When Fish tested two scaled-down models of humpback fins, one with tubercles and one without, he found that the tubercle-laden one performed much better in the wind tunnel. In particular, it could still generate lift at angles 40 percent steeper than the smooth fin. At those angles the smooth fin stalls out, like an airplane. For the whale, having bumps on its fins means it can roll over on its side that much farther

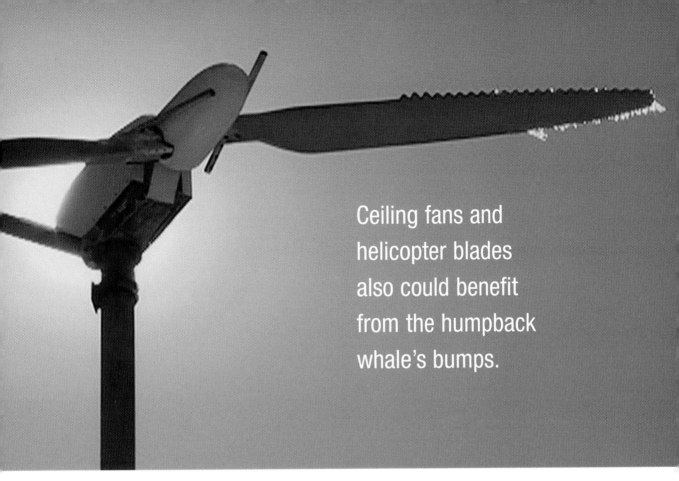

Ceiling fans and helicopter blades also could benefit from the humpback whale's bumps.

▲ The scalloped edges of Whalepower's fins seem to give them an aerodynamic advantage over smooth fins.

and still be getting a powerful grip on the water, meaning tighter turns and, ultimately, more food. Fish is pretty confident he knows what's going on.

"What happens is that you get some flow going over the bumps, but then to the side of each bump, you have a sort of valley. And what happens is the flow of water gets channelled into that valley and, as a result, the water then creates a vortex, a swirling mass over the surface of the flipper. It helps to accelerate the flow over the bump itself, but then it also keeps the flow of water attached to the surface of that flipper. That means more lift and less drag."

You can see what happens when the fins are tested in a wind tunnel. As the angle at which the fin meets the flow of air (or water) grows, more and more tufts start to flutter, showing that the airflow over the fin is being disrupted. Having tubercles delays that process significantly.

That's the whale story. The blades of a wind turbine are in roughly the same position as the whale's fins: they must "grip" the wind and generate enough force to turn the

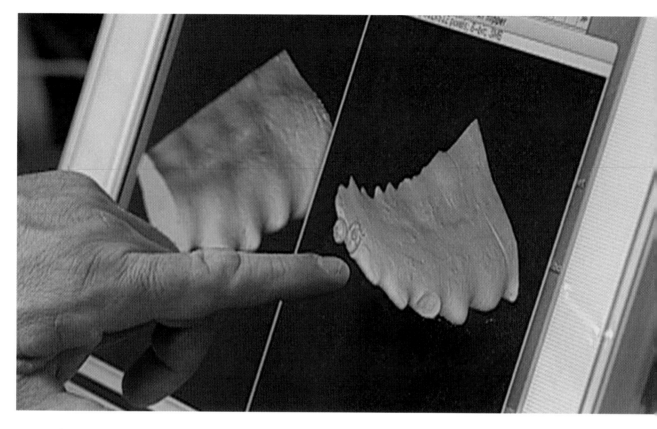

▲ Computer models of humpback fins suggest how turbine blades should be built.

turbine and generate electricity. But wind turbine blades are almost always smooth. Here's where Whalepower Technology enters the picture.

This Toronto company had the bright idea of building tubercles onto the leading edge of wind turbine blades to see if they would experience the same sort of effect: less drag, more lift. Whalepower has tested its first blades at the Wind Energy Institute of Canada in Prince Edward Island, and the results have been promising: there's not much difference at low speeds, but at medium winds and higher, the blade seems to generate more power, in some circumstances getting as much power from a 15-kilometre-per-hour wind as ordinary blades get from a wind blowing at more than 25 kilometres per hour. Whalepower's blades have also proven to be more stable, quiet, and durable, resisting even good Canadian wind-driven snow and ice.

There's no reason this technology should be restricted to wind turbine blades; ceiling fans and helicopter blades also could benefit from the humpback whale's bumps.

▼ The only way to mimic the fin accurately is to use fins from dead whales that have washed up on shore.

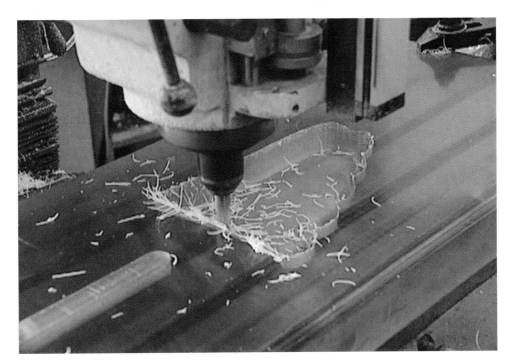

BIOFUELS:
TAKING CARS ON A ROCKY ROAD

THIS IS AN EXAMPLE OF A FOSSIL-fuel substitute that is generating heat before it even enters a gas tank. Years ago, hope for biofuels ran high: they were seen to be a substantial part of the solution to growing carbon emissions. If cars, trucks, trains, and even planes were to use fuel—specifically ethanol—derived from plants such as corn, the CO_2 problem would immediately lessen. It's not that these fuels don't release CO_2 when they're burned—they do. But that CO_2 was taken from the atmosphere by the plants just months before, so there's no net gain. Fossil fuels, on the other hand, release CO_2 that up until now has been locked away underground. In effect, they emit "new" CO_2 when they burn.

What could be better than biofuels? Brazil was already showing the way with a booming biofuel industry built on converting sugar cane to ethanol, and countries in the northern hemisphere committed themselves to becoming quick adopters. But as

▼ Corn is pulverized before it is fermented to produce ethanol (left). Excess solid material (right), left over from the fermentation, is made into high-grade animal feed. It all sounds too good to be true, and perhaps it is.

There is enough wind to power the whole world seven times over with wind energy.

more and more experts analyzed the case being made for ethanol in particular, and biofuels generally, flaws appeared. Some analyses claimed that the greenhouse gases emitted in growing, harvesting, and fermenting the corn sugar into ethanol produce so much CO_2 that the process is worthless, even *doubling* greenhouse emissions for thirty years, according to one analysis; others have argued that there is still a benefit, although it is not as dramatic as previously claimed.

The crop being grown is crucial as well. It has become clear that diverting corn from the food supply to the fuel supply has more than one downside: it can create food shortages, drive up the price of corn, and encourage farmers to clear forested land to grow more corn, and by doing so remove trees that would absorb, not generate, CO_2. That same argument can be applied to Brazil's thriving sugar cane–to-ethanol industry, where tropical forests are being converted in the same way. In fact, rather than using crops that have been planted on recently cleared land, the better choice by far is to use nonfood crops, like switchgrass, or crop waste and city garbage: stuff that you would walk on, not eat, material that might be cleared and burned, just to get it out of the way. And this stuff wouldn't be grown on land that, say, used to be reserved for corn, because the very act of converting the land creates more CO_2.

The list of problems goes on and on: one study showed that crops, such as corn, that need nitrogen fertilizer will emit copious quantities of nitrous oxide when combusted. Nitrous oxide is a potent greenhouse gas, and it just might be worse for global warming to burn materials that emit it than it would be to stick to fossil fuels! Another study showed that if both greenhouse gas emissions and effects on the environment, including human health, are calculated, the major biofuels—Brazilian ethanol from sugar cane, North American ethanol from corn, and diesel from Malaysian palm oil—are *all* worse than fossil fuels.

That biofuels would affect human health may come as a surprise, but in March 2007, *Daily Planet* sought out professor Mark Jacobson of the Department of Civil and Environmental Engineering at Stanford University. He had just completed a study looking at the possible health effects of biofuels.

"We did a study looking at the effects of converting all the vehicles in the U.S. from gasoline to ethanol. And what we found was that such a conversion would increase the death rate from gasoline vehicles—which is already ten thousand per year in the United States—by another two hundred per year, which is the size of a really small town."

Jacobson was looking at a fuel called E85: 85 percent ethanol, 15 percent gasoline. He found that cancer rates would likely stay about the same as they are with gasoline, but there was a significant difference when it came to ozone. Jacobson's projections found that in some parts of the United States, particularly Los Angeles, the use of E85 significantly increased levels of ozone, which is an important component of smog. Ozone can worsen asthma and have other effects on lung capacity and the immune system.

Jacobson has become convinced that the whole idea of biofuels is a nonstarter, and that we have to look further into the future and make the decision that electric cars powered by the wind are what we need.

"There is enough wind to power the whole world seven times over with wind energy. And that's not only electricity, but all the energy needs including vehicles, industrial requirements, and households. Another reason I like wind power is that it doesn't require nearly as much land area as other fuels like biofuels. And it's a nicely distributed energy source. It's not like a nuclear power plant which could be a target for a terrorist attack."

▲ Mark Jacobson, environmental engineer at Stanford University.

So what is the future for corn-based ethanol? While it may represent only a marginal improvement over gasoline from the global warming point of view (and maybe not even that), the truly promising biofuel is "cellulosic" biofuel, fuel made from waste materials like wood chips and switchgrass. In the United States alone there's a billion tonnes of the stuff. But the cellulose in "cellulosic" is the problem. All of these materials contain it, but so far it's way too expensive to break it down and prepare it for the gas tank. And the hopes for the discovery of some sort of bacterial superbug that can break down cellulose cheaply is, so far, just that: a hope. Corn isn't ideal, but at the moment, it's a lot cheaper than any of the alternatives.

FOUR
Getting Around

Transportation is an inviting target for climate change activists: ships are bad, cars and trucks are worse, airplanes the worst of all. Amid the controversy, a few free-thinking individuals have turned things around. Instead of trying to minimize the amount of energy consumed by various modes of travel, they're actually designing devices that enable us to *generate* energy. All we have to do is put one foot in front of the other.

HARVESTING ENERGY

THERE'S NO FREE LUNCH, BUT THERE IS FREE ENERGY, OR THE NEXT best thing. For instance, a Swedish company, Jernhusen AB, is planning to capture the body heat from the two hundred thousand or so daily commuters walking through Stockholm's Central Station, and use it to warm nearby buildings. Researchers figure all they have to do is install a few pipes to carry the water that will absorb the commuters' thermal energy, and they might offset 15 percent of the heating in a 40,000-square-foot building. And the people walking through the station won't even notice.

Also, scientists at the University of Pennsylvania have designed a backpack that generates energy from the up-and-down motion of walking. You may not realize that you bob as you walk, but you only have to think of videos of crowded streets with pedestrians walking toward the camera to realize that walking involves a lot more than just moving smoothly forward. In demonstration experiments, backpacks with a range of loads, from 20 to nearly 40 kilograms, generated power ranging from about 3 to over 7 watts, more than enough electricity to power an MP3 player, a BlackBerry, night-vision goggles, a handheld GPS, a Bluetooth wireless transceiver, and a couple of cell phones. Researchers are now working on adding a refrigeration system to keep vaccines cold while they're being delivered to remote areas, and they have since boosted the output to 20 watts. Compare that power output to the best ever generated by a device that derives electricity from the pres-

▶ This backpack frame generates energy from the up-and-down motion of normal human walking.

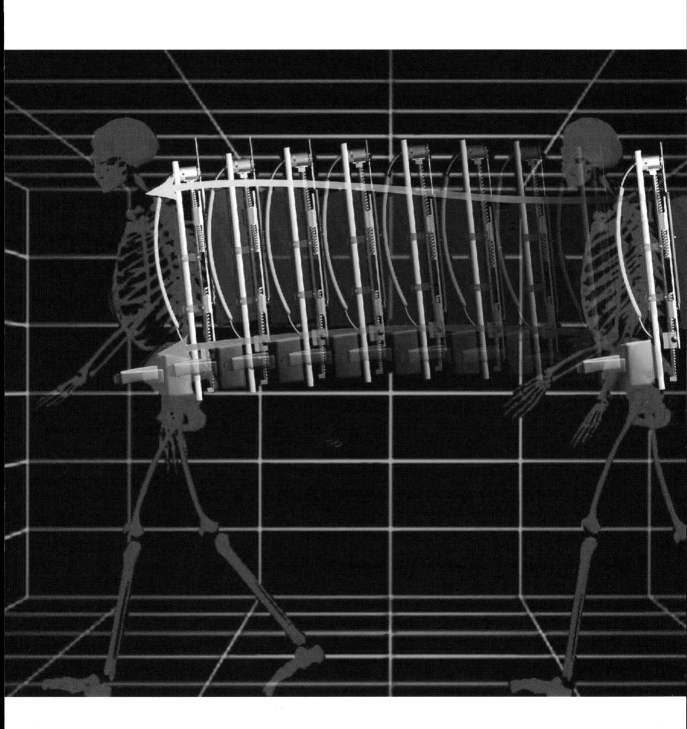

sure of a foot on a shoe while walking—a few thousandths of a watt. The backpack demands surprisingly little from the wearer: some mechanism, which still hasn't been identified, kicks in to minimize the additional energy you have to put out when you're wearing a backpack.

In the same spirit, Max Donelan and his students at Simon Fraser University have developed the Biomechanical Energy Harvester. It turns muscle power into electric power, and does so with surprisingly little impact on the person whose muscles are being used.

"We can get 5 watts. Now, for context, 5 watts is enough to charge ten cell phones simultaneously. So that's a lot of power. And then if you remove the restriction and allow people's effort to go up by a modest amount, we can get as much as 13 watts, nearly enough to charge thirty cell phones at the same time. Or put it another way, you can get thirty minutes of talk time for one minute of walking."

The Energy Harvester is fitted into a specially adapted knee brace, and weighs about 1.5 kilograms. Here's how it works: the muscles in your legs are always active when you're walking, but they're not always devoted to moving you forward. As your leading leg straightens out in front of you, before hitting the ground, your muscles put the brakes on, decelerating your leg so that it doesn't crash into the ground. That braking action can be turned into electricity, in the same way as a hybrid car converts energy normally lost in braking into charging the battery.

"We essentially do the same thing. We use a generator to assist the muscles in slowing down the leg at the periods of time during the walking cycle when the muscles are acting like brakes. We use a sensor to measure knee angle and a control system to determine the right stride time of the walking cycle to allow the generator to generate power. And in doing so, we can produce substantial electricity without increasing the effort required by the user."

Because the generator is not only converting muscle power to electricity but helping the deceleration phase of every step, volunteers who have worn the Harvester actually need to adjust when it's taken off: "Once we take it away, they now don't have enough braking force at this particular point in time, and so for a few strides,

"People are an excellent source of portable power. An average-sized person stores as much energy in fat as a 1000-kilogram battery. People recharge their body batteries with food and, lucky for us, there is about as much useful energy in a 35-gram granola bar as in a 3.5-kilogram lithium-ion battery."

MAX DONELAN

freq — one step delay.
tm accel — ∞
tm delay — 0

freq = 1 step.
tm accel — low
tm delay — 0

f

Calibration

Torq

x × y

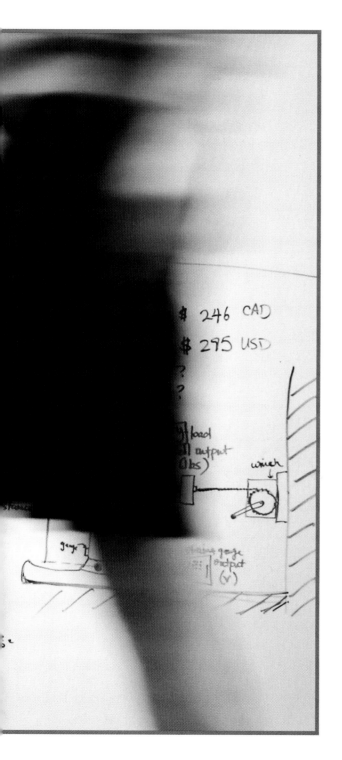

three or four or five, they're swinging their legs a little faster than they would normally. So they relearn to turn off the muscle activity again."

Donelan has also tried using the entire stride to generate electricity: he gets 25 percent more power that way, but the walker has to exert more energy to maintain speed. But regardless of which approach is used, it's energy with very little or even no cost. Donelan envisions this device being used by people who really need electrical energy on the move, like the military.

"Portable electricity represents much more than just a convenience to them. It allows a soldier to communicate, navigate, and get home safely. They would like to be able to use energy harvesting to lighten their battery load. They can sometimes carry as much as 13 kilograms of batteries for a mission. That's 30 pounds. So if instead they could carry a much smaller number of batteries and continuously recharge them with their own motion, it would allow them to use more sophisticated equipment. They could be the juice."

Donelan has no doubt that this early version of the Energy Harvester can be lightened and made more comfortable, so who knows? One day we might go for a walk and charge our cell phones at the same time.

◀ Max Donelan's Energy Harvester shows how even one piece of the jigsaw puzzle to combat energy consumption and greenhouse emissions can be important.

THE PROPHETS

> "The cookies are all reusable as well. These cookies we bring with us daily and what we do is we chew and suck on them for a while and then we pull them off our tongue and form them back into cookie-shaped cakes and put them back."
>
> ED ROBERTSON

"Totally New to Rock and Roll"

The Barenaked Ladies

The Barenaked Ladies have come a long way since they were removed from the bill for the 1991 New Year's Eve concert at Toronto's Nathan Phillips Square because the mayor thought their name objectified women. Actually, the reaction against the decision might have helped propel their career. Certainly today there's no doubt they are one of Canada's best-loved bands.

But they are aware of the wider world beyond the concert stage and the recording studio. *Daily Planet* caught up with them in the middle of their 2006/2007 tour of the United States and Canada, a tour on which they did everything they could to minimize the negative impact on the environment, with special attention to their carbon dioxide emissions.

The buses they used to travel from one concert venue to the next were running on 20 percent biodiesel fuel. The band calculated this one step kept 27 tonnes of CO_2 out of the atmosphere. To put that into perspective, the Canadian average per person, per year is just over 5 tonnes. It wasn't a perfect solution, but it was a start. Band member Ed Robertson notes that "in Canada it's tough to get distribution for the

▶ Foot-stomping music, but a minimal footprint.

fuel. That's the biggest hurdle for us. It was a lot easier in the U.S.; the network's broader." In fact, Steven Page recalls that in "western Canada, we had to go without for most of that trip because we couldn't find biodiesel."

The band tried to examine every detail of the tour to make the entire package as earth-friendly as possible. "You can go to a company like Bullfrog Power or Zerofootprint and you tell them your travel plans," Steven Page explains. "Bullfrog Power are offsetting electricity we use during the shows. Zerofootprint are offsetting our travel emissions. So for instance they might calculate over the course of a two-hour show we emit x-pounds of CO_2. So we give

them an amount of money to pay for the equivalent amount of green energy to put back into the grid."

Carbon offsets of the kind Steven Page mentions are big business now. The idea is to calculate how much carbon dioxide you're likely to emit, in this case, doing a concert tour. That in itself is a tricky calculation, involving the generation of electricity at every site, travel, even emissions accumulated during hotel stays. To offset these, a company like the Canadian nonprofit Zerofootprint will plan a project, for instance, tree planting to ensure that an equivalent amount of CO_2 will be absorbed by those trees.

But a legitimate carbon offset project takes the entire environment into account. It's not just a matter of planting a hundred trees in the bare ground left behind by a clear-cut. In fact, that is technically reforestation, and it's required by law wherever a natural forest is logged. Instead, carbon offsetters aim to *restore* a degraded or altered forest to its original state. This can mean planting native species of trees and managing them until they are established. By doing so, the normal process of the gradual succession of species is accelerated

◀ Even fans are asked to play their part.

dramatically. Since this is then a forest that wouldn't have regenerated in this way naturally, it qualifies as a genuine offset.

Offsets have their critics, however, even within the environmental movement. Their argument is that buying an offset simply assuages your guilt and allows you to continue pouring carbon dioxide into the air, rather than cutting down on your emissions. It's also true that *caveat emptor* presides: some companies claiming to provide offsets have dubious credentials; sometimes planted trees die. However, at least in these early stages of the struggle to contain CO_2, it's pretty clear that offsets represent a way of not only at least partially coping with the problem but, probably more

important, getting the word out—putting the idea of acknowledging and compensating for your carbon emissions on the map.

The Barenaked Ladies took other less dramatic steps toward making their tours eco-friendly, like encouraging concert-goers to buy their known carbon offsets to cover the expense they incurred getting to the concert, and even going so far as to change the entrenched backstage practices of using paper plates and Styrofoam cups. You might not think switching to china and glass is much of an environmental statement, but the thing it promotes, above all, is *consistency*. In the end, we can't afford to act responsibly only some of the time.

You can rock *and* be good to the earth.

PATRICK'S PLANK ON WHEELS

SKATEBOARDS ARE DEFINITELY ONE of the coolest forms of personal transportation. And what could be more environmentally responsible than a simple board with wheels that is totally propelled by human energy? Well, a lot actually. The materials that skateboards traditionally have been made of are not the most environmentally benign. Most contain formaldehyde (technically a carcinogen), and many of them, like carbon fibre, are derived from petroleum products, linking them to the use of fossil fuels.

When *Daily Planet* dropped in on Patrick Govang and Jason Salfi in 2007, they were collaborating on building a new kind of skateboard, made from recyclable material and designed to be completely biodegradable. "Ever since 1998, when we first started experimenting with different materials, I was ultimately hoping to find a bio-based replacement for carbon fibre epoxy in fibreglass," Govang explains, "and

finally in 2004 I came across something that mimicked the properties pretty well."

Govang and Salfi were trying to replace the traditional materials with fibres from the kenaf plant, a tall, reedy plant that has traditionally been used in Asia and Africa for rope and paper. They wanted everything that went into the skateboard to be grown in one year. Besides plant fibres, they experimented with alternatives to the traditional resins.

"The resin [we're using] is based on soy protein or soy flour, which today is used primarily in animal feeds," notes Govang. "And we're working in New York State with a biodiesel group that is using the oil from soy. We take the byproduct of that, the protein that comes off, and process that with water to create this resin. So if we're lucky this thing is going to cross-link and we're going to have a really strong

▶ Patrick Govang's skateboards are completely biodegradable … and practically edible too!

144

▲ Patrick Govang's biodegradable skateboards make it possible to be cool and earth-responsible at the same time.

skateboard made of 100 percent biodegradable material—cutting Johnny Oil Man out of the picture."

In fact, when our cameras were there, things didn't work out perfectly. The board was flexing, and it wasn't breaking, but then it flexed once too often—and broke. But hey, once you've gone that far, you can't back off. They changed the formula, adding one more layer to make a total of five: two bamboo, two natural fibre, and one maple. And this time the board withstood even the best that riders could give it. In fact, the materials they're making now have fabulous strength-to-weight ratios.

Since those early days (2007!), Comet Skateboards, Salfi's company, has launched twelve models of natural skateboards. Govang's company, e2e Materials, is looking to expand from skateboards to office furniture. The environmental benefits are almost too numerous to count: everything they make is biodegradable (no landfill here), no formaldehyde is involved, and there is zero reliance on fossil fuels. While you might think the amount of oil that is refined into the kind of products used in skateboards is trivial in the face of our total consumption of fossil fuel, if we are to change, we will have to change across the board.

"We're going to have a really strong skateboard made of 100 percent biodegradable material—cutting Johnny Oil Man out of the picture."

PATRICK GOVANG

Smog Veil

Lisa and Frank Mauceri

"The old building at 1825 West Wabansia has just been renovated. It's more than one hundred years old, but now it's state-of-the-art." How many times do you read something like that and just roll your eyes because you've heard it so many times before? But in this instance, a glance at the floor of that building in Chicago would make you realize that this isn't just any

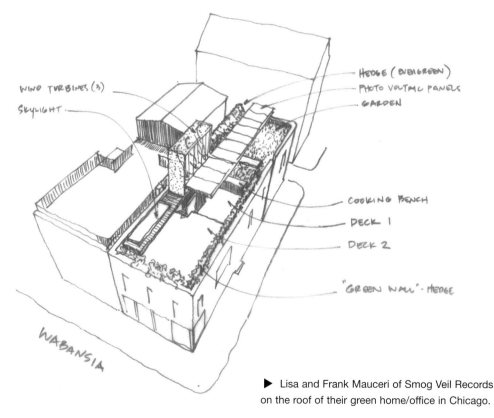

▶ Lisa and Frank Mauceri of Smog Veil Records on the roof of their green home/office in Chicago.

149

green story. The terrazzo floor is made from recycled glass and crushed vinyl from old records. The records were the out-of-date inventory of Smog Veil Records, the Chicago record label that specializes in three kinds of music: "innovative and experimental," "hideously rare," and "seriously ridiculous rock 'n' roll." The company is run by Lisa and Frank Mauceri. This is their office. It is also their home.

The Wis Tavern building (the tavern was in it for forty years) was constructed in the late 1880s. When the Mauceris decided to move into the building, and bring their Smog Veil label with them, they were determined to make this building as green as they possibly could. And they have.

This is a LEED-qualified building. The Leadership in Energy and Environmental Design (LEED) Green Building Rating System is the accepted measure of the greenness of a building. Most LEED buildings are factories, offices, or condos. The features that qualify this live/work space for LEED status include the floor with the old records (the ultimate remainder bin); the three wind turbines and thirty solar panels on the roof, which supply 50 percent of the building's electricity; 80 percent recycling of the materials resulting from the demolition of the original interior of the building (for instance, the original roof joists were used

▲ ▶ Cool and earth-friendly living in the heart of Chicago.

151

OFF THE BEATEN TRACK
(continued)

to make staircases); and geothermal heating and cooling in the basement, although it would be hard to see how this building and its owners could be any cooler.

As Frank puts it, "If one of the bands on our label decided to do an impromptu session in our office, they could plug in and do it powered off wind and sunshine. That's pretty crazy to think about."

Combining wind, solar, and geothermal is one of the things that makes this building unusual. There are fifteen geothermal wells in the basement, each extending nearly 20 metres down, drilled with a special undersized drill that could be used inside the house while it was being renovated; there was no room for a large drill on this urban property. In the summer, fluid

▲ Wind turbines and solar panels complete the picture.

warmed by the Chicago heat circulates down to where it's always about 13 degrees Celsius and cools before returning to the house; vice versa in the winter.

The wind turbines will generate 800 watts (as long as the wind is blowing about 14 kilometres an hour), the photovoltaics another 4800 watts. Total for the year should be about 10,000 kilowatts, or about half of their energy demands.

But it's not just about the building. The Mauceris want to change their business, album making and distribution: "The products we produce are energy-intensive and not full of recyclables," notes Frank. The plan includes ensuring that new releases are packaged only in 100 percent paperboard made from recycled materials and labelled with nontoxic inks (no more jewel cases) en route to an eventual 100 percent digital distribution of music. They are even targeting the typical promotional materials that accompany new recordings, like posters, promotional releases, and bios, aiming to provide reviewers with password-protected sites from which they can download any information they want (although I'm sure that printing that information will be discouraged).

Finally, and most important, the Mauceris want their bands to get involved by providing digital download gift cards instead of CDs at their gigs, and fuelling their buses with biodiesel.

How can you not like a woman who pushed architects and consultants to make their building as close to 100 percent sustainable as possible, but who's really proud of being an aficionado of the Frogger and Burger Time videogames, and says, "one of my greatest talents is my ability to beat Frank at those and foosball"?

How can you not like a guy who can say, at one moment, "Lisa and I are quite disturbed as to what lengths our country will go in the war for oil," but at another (remembering a time when he thought the music industry was foundering), "That's why I first signed the Spudmonsters"?

And how can you not admire what they're doing?

> "If one of the bands on our label decided to do an impromptu session in our office, they could plug in and do it powered off wind and sunshine."
>
> FRANK MAUCERI

A BICYCLE
POWERED BY NUCLEAR
FUSION—ON THE SUN

IN THE END, SOLAR ENERGY IN ITS VARIOUS FORMS COULD BE OUR SALVATION. Remember, forty minutes' worth of sunlight striking the earth contains as much energy as the entire world uses in a year. But how to capture that energy—that's the question. At the moment, photovoltaic cells are pricey; massive solar arrays like the ones scientists are researching at Sandia National Laboratories are still in the prototype stage, and nights and cloudy days remain issues as well.

However, it's not just up to the scientists and engineers to figure out solar. Sometimes it just takes an inventive mind, and a judicious mix of stubbornness and patience. Meet Peter Sandler.

"It's a very simple process. The sun's rays shine upon the solar panels, the energy they absorb from the sun is transferred into the batteries. The batteries in turn transfer the power

◀ ▼ Frank Sandler had the idea for a solar-powered bicycle long before the right technology existed to make it possible.

▼ Solar panels convert sunlight to electricity, powering the electric motor in the bike.

◀ The flexible solar panel is the secret to Sandler's solar-powered bike.

"I thought putting the solar panels in the wheels would make a very cool type of bike to ride around the city."
PETER SANDLER

to the motors. When you turn the key on and turn the throttle you have immediate power. That's how the bike works."

Yes, it's a simple process, but that doesn't tell the whole story—far from it. Sandler had the idea for a solar-powered bicycle at least ten years before it became a reality. As is often the case, the problem wasn't so much that his idea was impractical but that the right technology just hadn't come along yet. And that right technology, as Sandler explains, was flexible solar panels: "Basically if it was not for the flexible solar panel there would not be a solar electric bicycle. When I discovered that they did exist, I realized this was the opportunity to install them on a bike and that's how it all happened."

If you're like me, you might still be mystified about how flexible solar panels are the secret to the solar bike. But Sandler has no such uncertainty.

"I looked at the racing bikes that were used in the Olympics and noticed that they had those black disc inserts. So I thought putting the solar panels in the wheels would make a very cool type of bike to ride around the city. Many other attempts have been made to develop solar electric bicycles, but no one has attempted to do it in the framework of the bike itself, and that is what makes this bike unique."

Is the solar bicycle revolutionary? Well, you be the judge: it can reach 30 kilometres an hour, and you can imagine that at the end of a very long bike ride, a little assistance from the sun would be welcome. And you might need it: the bike weighs nearly 35 kilograms (75 pounds). I guess the bottom line is that if you're planning a long ride, make it a sunny day. But while the solar bicycle might not be revolutionary in that sense, it is a perfect example of how many, many ways of getting free energy there are, and in the end, that's what it's all about.

"We finally are beginning to realize that the earth has a voice."
BERNIE KRAUSE

The Sounds of Silence

Bernie Krause

Some of the most dramatic changes wrought by global warming are visual, the gradual disappearance of the world's glaciers being probably the most vivid. Sometimes the visual evidence of climate change is an absence, not a presence: already some North American birds are declining precipitously, although it's still too early to know how much of that decline is related to climate change. But what we sometimes forget is that a climate-changed world will *sound* different too. It's easy to forget how much of your daily environment you hear; those sounds are there even if you're only barely conscious of them. One thing is certain—you will notice when they're gone.

All of this is old news to a California man named Bernie Krause. For forty years Krause has been recording sounds all over the world—he now has nearly thirty thousand hours' worth of recordings! Over that time he has seen … well, heard … the world change.

"We finally are beginning to realize that the earth has a voice. It's critical. Without it, it's like looking at a silent movie. We can photograph the world from top to bottom but we don't really understand it unless we hear its voice. Don't forget that we have always had the ability to be able to capture

the visual, whether by painting on the caves in Lascaux, France, or by somehow graphically depicting the world around us. But we didn't have the technology to do that with sound until about 150 years ago."

Krause appears to be trying to make up for the delay all by himself. But there's much more than simply sheer volume to his recordings. Because he is careful to note the exact time of day and geographic location every time he records, he is able to compare past with present. Sometimes, the results are surprising—and disturbing: "Forty percent of my library is from places that are now extinct habitats, acoustically. And that's in forty years. I listen to some of these places and it brings tears to my eyes because … they are no longer there."

One of Krause's most striking examples is a pair of recordings from Lincoln Meadow in the Sierra Nevada. Unfortunately, we can show you only the sonograms of these sounds. The first, which demonstrates a rich variety of bird, insect, and other sounds, Krause recorded in 1988, just prior to the area being selectively logged. The logging company gave assurances that logging selectively would not alter the area enough to change what Krause calls the "biophony," or sound of nature. But the second sonogram, recorded a year later, makes it clear that the promise was empty—as empty as the recording itself. With the exception of a single Williamson's

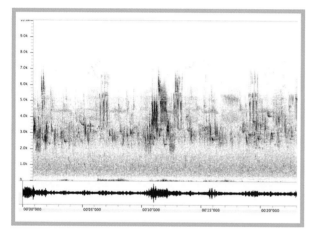

◄ The sonic landscape in Lincoln Meadow before logging.

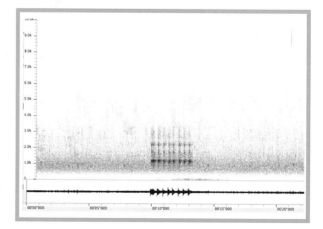

► The sonic landscape in Lincoln Meadow a year later. Even today, twenty-plus years on, the original soundscape has not returned.

Sapsucker hammering on a tree, Krause found no evidence of the natural habitat that existed there only a year before: "It's changed incredibly and all of the sound, all of the richness, all of the diversity has gone. And it hasn't come back. Yet if you photograph that, it looks perfect."

That was the impact of selective logging, and at the moment it would be impossible to predict how the biophony of any habitat will change with global warming. But without doubt there *will* be change.

"It's changed incredibly and all of the sound,
all of the richness, all of the diversity has gone.
And it hasn't come back. Yet if you photograph
that, it looks perfect."

BERNIE KRAUSE

▲ Lincoln Meadow in the Sierra Nevada. It may look the same as it did before it was selectively logged, but it does not *sound* the same.

FIVE
Where on Earth
Will We Live?

Our homes are a large source of greenhouse gases, producing at least 80 megatonnes of carbon dioxide each year in Canada (1210 tonnes in the United States)—high compared to the rest of the world. Part of the solution has to come from home designers and builders, but there is room for individual innovation ...

GETTING TO ZERO
IN EDMONTON

▶ Two homes, one amazing house. The NetZero house built in Edmonton, Alberta, will, over the course of a year, give back as much to the electrical grid as it takes.

9926-87 ST.

Even during a winter night at 32 below, the Riverdale NetZero home needs the equivalent energy of only four toasters or six hair dryers to keep the inhabitants warm.

▲ Builder Peter Amerongen has shown that even in northern latitudes, houses can be deeply green.

EDMONTON, ALBERTA, IS AT 53 DEGREES north latitude. This northern location creates multiple challenges for anyone hoping to build a home that uses the minimum of energy. Winters are cold, for one thing, and as sunny as the Alberta capital is, generating solar power from a low-hanging sun in that cold winter is a marginal prospect. But that northerly disadvantage didn't stop Edmonton home builder Peter Amerongen and his team from designing and building North America's most northerly "net-zero" home.

Rather than being utterly disconnected from the electrical grid, this house has a give-and-take relationship with it. Sometimes at night and during the winter the house draws electricity, but over the course of a year it makes that up (and then some) during the day and the summer, so that at the end of the year it will likely have fed some energy back to the grid.

Even during a winter night at 32 below, the Riverdale NetZero home needs the equivalent energy of only four toasters or six hair dryers to keep the inhabitants warm. Rather than trying to keep consumption of natural gas for heating at a minimum, the designers went cold turkey and never even attached a gas line. They applied the term "sustainable" to everything they used, avoiding products that, however efficient and economical, were either hydrocarbon based or caused high levels of greenhouse gas emissions during production.

What should encourage all prospective homeowners is that none of the technology or designs used in the Riverdale NetZero home are ridiculously futuristic or impossibly expensive. The home is a perfect example of how dramatic reductions in environmental impact can be achieved, not by a single extravagant step but by myriad smaller ones. It also illustrates the importance of thinking about sustainability before the shovel hits the dirt. As George Monbiot, in *Heat: How to Stop the Planet from Burning*, has said of his own house in England,

▲ Rather than being utterly disconnected from the grid, this house has a give-and-take relationship with it.

Had [my builder] spent an extra £1,000, he would have cut my gas bills in half. Fitting the roof insulation properly would have cost him next to nothing. Solid wall insulation would have cost more, but part of the price could have been offset by using standard light fittings instead of the more expensive embedded ones. As he was ripping up the floors anyway, it would scarcely have hurt him to have rolled out a few strips of fibre … But if we were to do what he should have done, we would need to gut the house all over again … It would cost us something like £20,000 to put it right.

George Monbiot's builder obviously had nothing to do with the design and construction of the Riverdale NetZero house. Take just one facet of the design: the threefold use of solar:

• On the south-facing roof, a 30-square-metre array of photovoltaic cells generates electricity, 6200 kilowatt-hours per year's worth. The cells are angled at 53 degrees, partly to minimize snow cover, partly to compensate for the high latitude. They will generate enough electricity to contribute back to the grid.

• Just below the photovoltaics is the "active solar heating," a row of transparent panels (22 square metres' worth) that expose water-filled pipes to the sun. That heated water is sent to two basement tanks, one of 300-litre capacity for hot water use in the home, another giant one, holding 17,000 litres, which heats the air in the home.

• Finally, below the active solar heating panels are nearly 17 square metres of triple-glazed south-facing windows (on the north side of the house they are *quadruple*-glazed) that allow sunlight to warm the interior materials, like concrete countertops, but prevent loss of heat from the house.

The house has no air conditioning; the minimal cooling needed can be provided by a loop of running water hooked up to the air circulation system, and by opening windows or shading them, measures that might have to be beefed up in hotter, more humid climes. Throw in the usual suspects like energy-efficient appliances, low-flow shower heads, and compact fluorescent bulbs, and there you have it: off the grid in a city that is renowned, even in the face of global warming, for its bitterly cold winters.

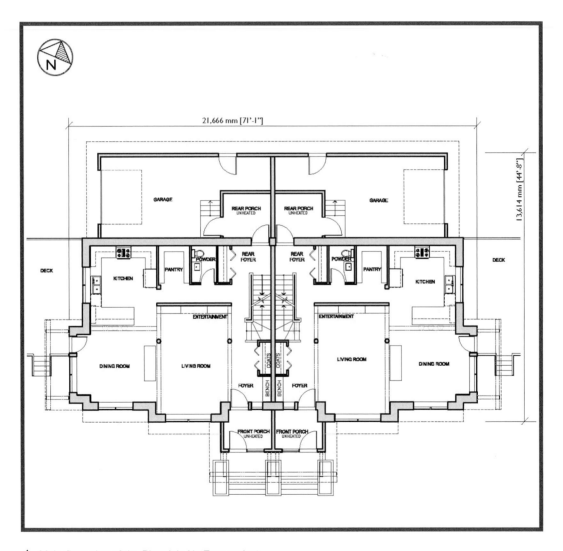

▲ Main-floor plan of the Riverdale NetZero project.

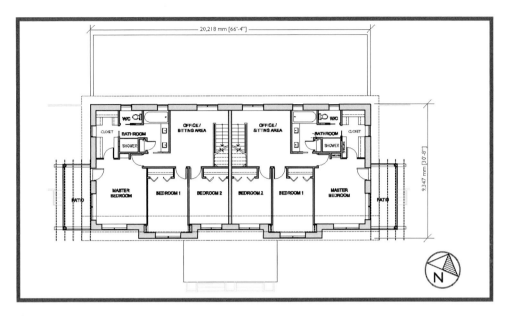

▲ The second-floor plan.

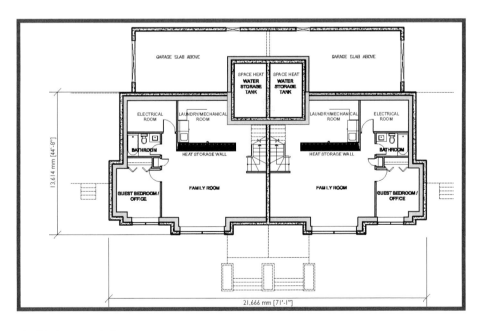

▲ The basement plan.

171

▲ Barbara Kerr was an environmentalist before it was the thing to do.

The Solar Granny

Barbara Kerr

Barbara Kerr is anything but a young solar energy activist. Even she would admit that in her mid-eighties, she's on the right-hand side of the curve. But so what? She's been plugged into the idea of living off the grid for a long, long time, and this house, all 960 square feet of it, has only a nodding acquaintance with the power lines 300 metres away.

Her house has all sorts of clever design features that make it a model of energy efficiency, but her solar oven is the coolest. You could call this the Mark VIII, the eighth version of Barbara's solar oven: it can reach temperatures of 230 degrees Celsius (450 degrees Fahrenheit). Today it's roasting chicken. But the Mark VIII wasn't easy to come by: "I'm not a natural builder and I had a lot of trouble sawing the angles and deciding when they said put a reflector at a 60-degree angle, did they mean 60 degrees from here, or from there? Just couldn't figure that one out."

▲ You too can have a solar cooker …

▼ … and cook with it.

The hundred-year-old windmill does its part at Barbara's place too.

Barbara explains that it all came down to experimentation: "I would put a quarter cup of rice and a half cup of water in two cans, and I'd put one in one stove and one in another. I'd wait and watch and the one that cooked the fastest was the stove."

In this house it seems that the principle of the solar cooker has driven the rest of the design. Jim Scott is responsible for some of those designs, and he took us on a tour: "This is a solar water distiller. This will distill over a gallon a day in the wintertime and about three in the summer when we get much more direct sunshine." And then there's the solar water heater: "Basically it is a large solar cooker with a water tank in it, an insulated box with a couple of inches of foam insulation and a tank painted black inside to act as a water collector."

There's much more to Barbara Kerr's house: the sun heats the insulated cement floor and the brick walls; the downspouts collect rainwater as backup for the water pumped out of the well. And the power for that pump? This hundred-year-old windmill. But in the end, somehow, it's all about the solar cooker.

"All over the world forests are going down because of the need for cooking fuel," Barbara explains. "We didn't realize at the beginning how serious it was, but I had a man who came from Mali and looked at the stoves and I didn't know what he was thinking, but he said, 'We've got to have these stoves, there's no wood left.' And that's true. It's not being frivolous to use wood like this—cooking is a basic need—unless we have some good alternative, and solar is the good alternative."

Barbara now has a company that sells prefab solar cookers. Reflective surfaces, a few clothespins, and a plastic bag, and you too can do chicken for dinner.

"All over the world forests are going down because of the need for cooking fuel."

BARBARA KERR

DOWNSIZING
PAR EXCELLENCE

TODAY'S AVERAGE AMERICAN HOME HAS FOUR TIMES THE LIVING space per person as it had in 1950. Let me be more specific: average floor space in 1950 was 983 square feet. By 1970 that had grown to 1500 square feet, and by 2004, ballooned once more to 2349 square feet. The size of every room (with the possible exception of the living room, which is going extinct) has increased. And that's just the floor space. When you consider that ceiling heights are also on the rise, the volume of air inside a home that must be heated or cooled is staggering. Surely this growth—accompanied as it is by *shrinking* families—should be halted, even reversed. Reversed? One enterprising Californian has done just that. Meet Jay Shafer and his Tumbleweed houses.

Shafer's houses range in size, but the ones pictured here are only about 100 square feet. That's 100 square feet, or roughly 10 by 10, smaller than most rooms. Way smaller than a bachelor apartment. But they're houses, and you have to agree that they are cool, although I'd worry they might appeal most to people who like a high degree of precision and order, like model railroad enthusiasts. In fact, most of the houses are indeed on wheels, so they can be delivered anywhere in the continental United States, for about $4 for every mile from Sebastopol, California, about an hour north of San Francisco.

▶ Jay Shafer and one of his Tumbleweed houses.

They grow on you if, in fact, they don't grow much themselves.

The greenness of this house is its size. It comes without any mechanisms for generating electricity, heat, or water. When your house is delivered, you have to plug into the local system, which could be a coal-fired power plant. But even if you burn carbon-emitting fuels, your footprint is as tiny as your house's: Shafer himself claims that he spent no more than $170 on propane to heat his little house over the winter in Iowa. This amount could even be reduced to double digits in slightly warmer areas. Of course, sustainable energy sources would make that bill look a lot better. Plugging into power and sewage systems makes the house function a little like an RV, but with many more of the comforts of home.

The house in the pictures, which is 110 square feet, costs about $46,000. You can buy a finished house at that price, or you can pay much less and just buy the plans for $1000, but then you need to buy the materials and find the construction crew to build it. Tumbleweed estimates you'd save a third of the costs doing it that way.

It's unsettling how the idea of a 100-square-foot house gets a grip on you. Go to Jay Shafer's webpage (www.tumbleweedhouses.com) and click on "houses"; if you're like me, you'll scroll down, see that the 75- and 100-square-foot versions are giving way to behemoths like the 400-square-foot Z-glass model, and wonder, "Why on earth would anyone want a house that big?"

EARTHSHIPS HAVE LANDED

THE HOUSES ON THESE PAGES LOOK A little like Hobbit dwellings, or the skyline of some outpost in a *Star Wars* episode. Nonetheless, these homes, called Earthships, are in the same vein as the others in this section: they all represent attempts to reduce energy consumption so dramatically that they can thrive off the grid, reduce their greenhouse gas emissions to the minimum, and still provide a comfortable life. Earthship homes go even further: they recycle grey water (water from dishwashing and showering) and even treat their own sewage, so they can be completely disconnected. An Earthship could exist anywhere.

Michael Reynolds created the company Earthship Biotecture. He is the "Garbage Warrior," a man who has known for decades how he wants to build environmentally friendly homes, and the secrets of doing that. His theme is use as little outside energy as possible; the two innovations that make that possible are incorporating thermal mass and making sure that most of it is composed of garbage. Stuff with high thermal mass stores great quantities of heat and releases it relatively easily. In this case, the garbage is old car tires, or beer cans, crammed full with some sort of dense material: the tires with soil pounded into them, the beer cans sealed and filled with water. When walls are built from these materials, and protected from the elements by being half-buried, the homes are like caves, varying their temperature relatively little from day to night, season to season. The result: almost no energy is consumed either heating or cooling the interior.

The conservation of water, something most urban dwellers ignore, is taken very seriously here. These homes can be hooked up to a water supply, but they are designed to collect all the precipitation that falls on the house, filter it, and use it for cooking, drinking, or bathing. Then it's reused: treated biologically in indoor planters, then piped to the bathrooms where it's used to

▶ Earthship homes are buried on three sides.

The homes are like caves, varying their temperature relatively little from day to night, season to season.

flush toilets. Finally, the resulting waste-water is given further treatment, either in a conventional septic field or in an outdoor botanical pond. If the pond is used, no wastewater is returned to the environment. As long as enough rain falls, the house neither imports fresh water nor exports sewage. This emphasis on water conservation may in the end turn out to be the most far-seeing of all the Earthship design innovations, especially as the climate warms.

Electricity is provided by photovoltaic panels and a small wind turbine. But these are details, important for the maintenance of the house, but not as eye-catching as the simultaneously futuristic and prehistoric look of the dwellings, and the astonishing sight of bananas being grown inside the greenhouse that is a standard part of the chain of water treatments. Those bananas grow in New Mexico ... could they be grown in northern Saskatchewan?

▲ Michael Reynolds, the "Garbage Warrior," inventor of Earthship houses.

▶ Nowhere on an Earthship is the fate of water left to chance.

THE PROPHETS

The Greenest CEO
Ray Anderson

Here's the difference between fame and worth: Ray Anderson gained a lot of fame through the documentary film *The Corporation*, but he had established his worth long before. In the movie, he was an anomaly, a corporate guy who had seen the light and was bent on turning his company, Interface Carpets, into something never seen before: a high-profit operation earning those profits sustainably. It wasn't just the shock of seeing a CEO in a suit acknowledge

▲ Carpet making: until now, a resource- and energy-intensive industry.

that climate change and carbon emissions were serious issues, it was the way he said it. He admitted to having been, through his company, a "plunderer," and foresaw a time when people like him would go to jail.

But all of this started much earlier than the movie, in the early 1990s, when Anderson realized that he was woefully ignorant of environmental issues. As I heard him tell it, some of his staff approached him to ask if he'd contribute a few remarks about the environment to the group, and he realized that he didn't actually know enough to say anything. (When Interface clients would say to sales reps, "Interface just doesn't get it," Anderson wondered, "Doesn't get what?") At about the same time, the book *Natural Capitalism* by Paul Hawken, Amory Lovins, and L. Hunter Lovins was on his desk. Anderson read it, had an epiphany, and now considers himself a recovering plunderer. And when you size up what Interface has done over the last few years, he's entitled to the adjective "recovering" and much more. Since Interface began to rethink its approach to the environment, it has

- reduced water usage by 75 percent
- eliminated 80 percent of its effluent pipes
- reduced the amount of waste going to landfill by 70 percent

▲ Ray Anderson has pledged that by 2020, Interface Carpets will use *no* new natural resources that are not sustainable.

Interface also has dramatically reduced its greenhouse gas emissions by using more renewable energy and using it more efficiently. The company collects landfill gas from a site in Louisiana and converts it to fuel, reducing both its consumption of fossil fuel and eliminating the greenhouse gas that would have been emitted by that fuel.

Of course, the bottom line is … the bottom line, and even here Anderson's company is exemplary: since 1996 its profits have doubled even as the market shrank by 36 percent.

Some of those profits he attributes to the "goodwill" of the marketplace. He is the inspiration behind Interface's Mission Zero: by 2020 the firm will "eliminate any negative impact the company might have on the environment," totally recycling, taking nothing from the earth and returning nothing to it—except perhaps energy. Interface aims to be not just sustainable, but *restorative*. And Anderson says at the halfway point, 2008, "we are on schedule."

Ray Anderson's approach to business and the environment is seen most clearly when he contrasts old and new ideas:

OLD: "Earth has an infinite supply of stuff." NEW: "Earth is finite." As Anderson has pointed out, industry processes four million pounds of material, from getting it out of the ground to wrapping it in plastic, to provide one average, middle-class American family their needs for a year, not something a finite earth can withstand.
OLD: "Earth is an infinite sink that will absorb whatever we dump into it."
NEW: "Earth is finite *and* fragile."
OLD: "Technology will see us through."
NEW: "Technology is part of the problem."
OLD: "Industrial productivity."
NEW: "Resource productivity, making our resources work for us many times over."
OLD: "The environment is a subset of the economy."
NEW: "The economy is a wholly owned subsidiary of the environment."

Here's an example of changed thinking: a typical Interface carpet tile, which contains six different polymers, can be anywhere from 38 to 94 percent recycled material, depending on things like the type of fibre and whether the backing itself is recycled. Recycled material is now so precious to the company that it's seriously thinking about mining landfills! *That* is thinking differently.

But even with the successes of his company, and even if he says, "I see barriers coming down everywhere, and we've come light years in a very short period of time," if you ask Anderson whether we'll right the wrongs in time, he might first say, "My researchers—and they are among the best in this field—tell me that not one peer-reviewed scientific paper published in the last thirty years has said, yes, the global trends are now positive." But then he will say, "Will we succeed? That is the question of our era."

AN ICONOCLAST AND HIS HOUSE

BILL LISHMAN'S AFFAIR WITH ULTRA-light aircraft tells you all you need to know about him. First attracted to them when he was teenager, he eventually became an accomplished pilot. He had also always wanted to fly with the birds—he knew about Konrad Lorenz's experiments with imprinting, and he put it all together: if he could get young, hand-reared geese to imprint on his ultralight, they'd follow him in the air as they would their mother. That's exactly what he did. He then switched to a much more significant bird, the whooping crane, and taught young whoopers to migrate south from a Wisconsin refuge to a new wintering ground in Florida. That new migratory route for the still critically endangered species was a triumph of nature, with the help of a little cool technology. That same capability for innovation led Lishman to build, well, a different house.

As he puts it, "Houses should be far more efficient and last much longer. Here in Canada a house lasts about a hundred years, but in Europe they double or triple that."

"I was living in an inefficient house and because it was a windy spot, I thought I could put up a wind generator," Lishman goes on to explain. "But the heat loss from the house turned out to be much more pressing. How could I design a more earth-friendly house? Going down into earth was the way to go, 'a buried igloo.'"

But going down into the earth isn't that straightforward. After bringing in earth-moving equipment to shear off the top of a hill, Lishman laid out his house, a set of onion domes linked together, with the largest, the living room, in the centre. Each has a skylight at the highest point of the room. Before the earth was moved back onto the structure, the exterior was covered in sand, and air ducts were laid down throughout to bring warm air from the sunny side of the house. Then a rubber sheet was put on top, with soil covering the sheet. Now the roof is a garden.

▼ There is minimal exposure to the elements.　　　　　▶ Even a buried room can have great sunlight.

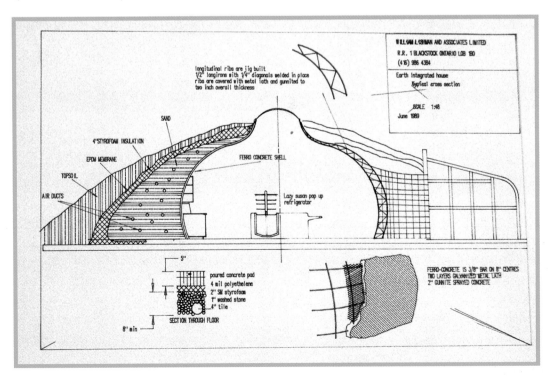

Lishman went for circular underground domes for a reason: "Most of the underground architecture of the time kept the rectangular shape. That's the wrong shape for a house both psychologically and technologically."

Psychologically? It's true that you'll spend plenty of time in an underground box after you die. Lishman wonders if this unpleasant association may be at work here. Even if not, the rectangle still has its drawbacks. For one thing, it would have to be a very strong box to support the enormous overburden of soil. Also, it's hard to light an underground rectangle: the corners are usually dark. But an onion-shaped house eliminates most of those drawbacks; it never has to be painted, shingled, or have the eavestroughs cleaned out; and the roof is covered with a garden or lawn.

"I built the house in 1989–90, and global warming wasn't the issue then that it is now; if anything, saving energy was the big thing. But it was clear that it wouldn't take too long before we were running out of fossil fuels."

You'd probably guess that a buried house doesn't rely much on solar, and you'd be right. The only solar energy Lishman uses is passive, but it is effective enough: it gives him an extra month and a half before he has to start heating the house in the winter, and he's usually able to stop a few weeks ahead of time in the spring. But solar aside, there's much less wear and tear on the house because it's buried. Even so, it's not completely immune to a Canadian winter: Lishman virtually closes down the rooms on the north side until spring. But for all the environmental advantages, in this house it's the aesthetics that reign: "I don't think we're meant to live in boxes," Lishman explains. "There's a feeling of space and light in this house. In a conventional house with a nine-foot ceiling it's constantly overcast. With a dome and a skylight it's like being under a clear sky."

▼ "There's a feeling of space and light in this house ... With a dome and a skylight it's like being under a clear sky."

▲ Bill Lishman's house started out as seven separate domes with skylights at the top.

▼ Today the domes are invisible, hidden under the roof garden.

"I don't think we're
meant to live in boxes."
BILL LISHMAN

THE PROPHETS

The Environmental Evangelist

Amory Lovins

Long before Ray Anderson of Interface Carpets had his epiphany, Amory Lovins was an environmental evangelist. Evangelist might not be the right word, because although Lovins expresses himself with fervour, he is not asking anyone to take things on faith. He has numbers, data, projections backing up all his claims.

Amory Lovins is chairman of the Rocky Mountain Institute in Snowmass, Colorado. His home/office could be featured in this section of the book, seeing as how it has no furnace even though it experiences below-zero winter temperatures, a set of solar panels on the roof tracks the sun, the windows do not let out heat in the winter, and clothes dry on an indoor line. He walks the talk, and sometimes even makes it up: he invented the word "negawatt" to mean a watt of electricity that does not have to be generated because an energy-saving measure has eliminated the need for it. And when he talks, he makes strong claims. He believes that North America can be weaned off its dependence on oil by 2050 without any dramatic technological changes, and without any significant change in living standards. "You can re-double the efficiency of using oil. We've already doubled it since '75, but we can double it again at an average cost of US$12 per saved barrel. Then we can replace the other half of the oil with a mixture of, say, natural gas and advanced biofuels, mainly ethanol made from woody/weedy stuff like switchgrass and forestry waste. That costs about US$18 a barrel, so the average cost of getting off oil is $15 a barrel." Lovins likes to quote the simple statistic that bears out his claims: from 1977 to 1985, the U.S. economy grew 27 percent while oil use fell 17 percent. But that was then—how could it be done again?

Amory: The key is tripling the efficiency of cars, trucks, and planes, and we can do that with present technology and get our money back in two years for the cars, one year for the heavy trucks, and four or five years with

▲ It looks like plastic, sounds like metal, and would reduce the weight of a car by half.

planes. For example, here's a material from a new manufacturing process that lets us make aerospace-grade carbon fibre composites at automotive costs and speed. So this is plastic, but it sounds like metal. It's actually incredibly strong and stiff. It will absorb twelve times the crash energy of steel per kilogram. It's tougher than titanium and we can squash this onto a hot dye to make this shape in about thirty to sixty seconds out of a flat blank that's made at 1.4 metres a second, out of a kind of ink-jet printer that makes the carbon fibres go where you want to get strength in the right direction.

Jay: And is it expensive?

Amory: The material costs more than steel, but the car costs the same. So if you made cars of this stuff you would lose half the weight and half of the fuel use, but the car would cost the same because the costlier material is offset by much simpler automaking and a two or three times smaller propulsion system. The car is so light and slippery it only takes a third of the power to move it.

There is a problem with efficiency that George Monbiot points out: the more efficiently we use something like oil, the more money we can spend on other uses for it (uses we now find attractive because it is cheaper), neatly and effectively negating the gains in efficiency. But Lovins is irrepressible. Where other environmentalists throw up their hands in despair at the thought of 1.3 billion Chinese and 1 billion Indians buying new cars, Lovins sees it as a "wonderful opportunity to do it better the first time."

Lovins, more than most, sees every facet of an environmental problem, and that's no different when it comes to the cars of the future—it's not just about cars.

"We need to get a lot better at having great cars and not needing to drive them so much, for example, best of all by being already where you want to be so you needn't go somewhere else. If you stop mandating and subsidizing sprawl, you have less of it. Or as a Canadian friend of mine said, 'Drivers should get what they pay for and

> ## "I don't do problems … I do solutions."
> ### AMORY LOVINS

pay for what they get.' We don't have the second part of that yet."

In an article in 1976, when very few were even thinking about global warming, Lovins wrote, "The commitment to a long-term coal economy many times the scale of today's makes the doubling of atmospheric carbon dioxide concentration early in the next century virtually unavoidable, with the prospect then or soon thereafter of substantial and perhaps irreversible changes in global climate."

He knew it then; he knows it now.

▶ Sheets of aerospace-grade carbon fibre composite material (the stuff of the next generation of cars?) being made in a matter of minutes.

ARCHITECTURE MIMICS LIFE

IMITATING NATURE IS AT LEAST PART OF the secret of the environmentally friendly houses designed by both Bill Lishman and Michael Reynolds. If part of the house is underground, wind exposure and the daily and seasonal ups and downs of temperature are minimized. It is a strategy used by countless animals and birds in Canada's North: dig under the snow and you're insulated from the cold. But there's another step that could be taken, and it too takes a page from nature.

Countless plants, especially in the Arctic, turn on their stems during the day to track the sun, both to stay warmer and to maximize the amount of solar energy for photosynthesis. Some flowers are actually shaped like solar-collecting dishes. If solar energy is to be part of the mix in limiting the amount of energy our homes consume, could we take a page from these plants and turn our houses? Well, the answer isn't just "yes, we can," but "yes, it's already been done."

Two houses built on opposite sides of the world are beautiful examples, especially when put together, of how a simple change of approach could lead to dramatic new ways of building houses. The more straightforward of the two is Luke Everingham's

house in Australia. It's different; in some ways it's radical; and Everingham built it on a simple suggestion from his wife: "She said, 'Oh wouldn't it be great to have a house that you could move,' and I just started to think, why not?"

In Everingham's case, energy consumption wasn't the only consideration and, in fact, wasn't even number one. He wanted something different from his former home:

▼ Luke Everingham built an eight-sided house with one remarkable property: it turns.

"We had to go outside to see the views or any wildlife or anything like that. And it was very dark inside. And in summer, it lacked air flow."

To correct these flaws, Everingham designed and built this eight-sided house, with 5 metres of floor-to-ceiling glass in nearly every room. That wouldn't work so well in a Canadian winter, but it's perfect down under, and the true genius of the place is found in the mechanics of rotation.

"There are two drive motors. They're situated 180 degrees apart, and there are thirty-two of these wheels, each one with a 5-tonne rating. The track has to be perfectly flat, otherwise the undulations would cause distortion in the framework above. The challenges were that everything's got to be perfect, unlike a normal house where a builder's happy to be 20 millimetres out

▼ It conserves energy by avoiding the Australian sun, and *you* choose the view.

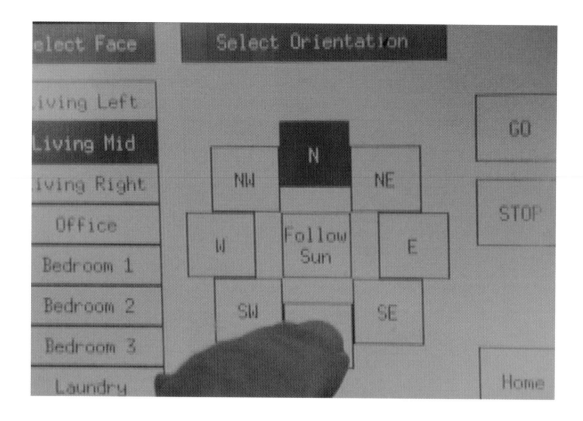

here or an inch there. This is absolutely perfect or it's a failure."

His house makes two full turns every hour, a speed that you don't really notice much once you're inside. But you might wonder why the wires and other contacts with the outside world don't just get twisted until they snap. The house has a central cylinder through which all such services enter and exit: a pipe 120 metres long brings geothermal heat in, and electrical wires are bundled together and super flexible, able to withstand 720 degrees' worth of turns, although Everingham's house turns only

360 degrees in one direction before reversing its field.

The rotation device is surprisingly easy to operate—and inexpensive too: "We just press, say, living room ... follow the sun ... go. There are 4800 reference points on the encoder and the computer knows 365 days a year at any time of day where the sun will be. We're using two 500-watt motors which are about two-thirds horsepower. They're about the size of a washing machine motor, and when you work it out, it's costing us about a dollar a week to rotate the house."

If you didn't know Luke Everingham's

"We just press, say, living room ... follow the sun ... go."
LUKE EVERINGHAM

▲ Rolf Disch has taken the rotating house several steps further, creating what he calls a "plusenergiehaus," a building that generates more energy than it consumes.

house rotated, you might not even take notice of it. It is eight-sided, but in most other respects it appears to be a typical house. On the other hand, in Freiburg, Germany, architect Rolf Disch also has built a rotating house, but this one is very different, a three-storey home where the environmental features, not the view, are totally up front. It turns in sync with the sun, either to face it in winter or to avoid it in summer.

Disch calls his house Heliotrop (the English version of the word adds an *e*); the word roughly translates as "sun-seeking" and is taken from those Arctic plants, which by tracking the sun can maintain temperatures in their flowers 7 or 8 degrees Celsius warmer than the air around them. Disch's eighteen-sided heliotrope uses exactly the same technique: gathering solar heat passively in the winter through its windows, but then turning them away from the hot summer sun. However, Heliotrop also turns solar energy into electricity using a free-turning photovoltaic array on the roof. The

array has to be able to turn independent of the house because it must generate electricity—and be turned toward the sun—even on the hottest days of summer when the windows are facing north. It produces something like five times the electricity needed to run the house.

Tubes winding around the building collect solar energy to provide hot water (and simultaneously serve as railings for balconies). Throw in some geothermal and you have a house that generates its own energy. A spiral staircase winding up the axis of the house provides access to all the living spaces inside.

It is a fantastic design, so completely radical in appearance that you can be fooled into thinking it must be a one-off, something that only a free-thinking and obsessive architect could come up with. And maybe it is. But Heliotrop and Luke Everingham's house have one thing in common. They make you wonder whether rotating houses might actually make sense.

The house makes two full turns every hour.

THE TRIUMPH OF DEATH

Pieter Bruegel the Elder painted *The Triumph of Death* in the mid-1500s. It depicts the horror of the Black Death, the epidemic of bubonic plague that swept through the Middle East and Europe in the 1300s. It's estimated that the disease might have killed seventy-five million people, including as much as 50 percent of Europe's population. As the Bruegel painting illustrates so vividly, civilization was in chaos.

But what does this have to do with global warming? According to William Ruddiman at the University of Virginia, the Black Death was one of several events in human history that altered the concentration of carbon dioxide in the atmosphere and changed the climate as a result. In fact, where others argue that human influence began in the mid-nineteenth century, and call this the beginning of the "Anthro-pocene" era, Ruddiman would push that date back several thousand years.

Ruddiman takes as his starting point ice core records that show that CO_2 started to rise, albeit slowly, thousands of years ago, with occasional pauses along the way. The connection to humans is farming, which began about eight thousand years ago. Ruddiman is confident that cutting down forests to cultivate the land released enough carbon dioxide to register in the ice cores.

▶ *The Triumph of Death,* by Pieter Bruegel the Elder.

Methane, another greenhouse gas, is part of this story too. About five thousand years ago, humans started to cultivate rice, and at about that same time, as seen in ice cores, the levels of methane gas in the atmosphere, rather than declining slowly as they should have based on the previous thirty-five thousand years' worth of records, stayed level or actually started to rise. Methane is a more powerful greenhouse gas than even carbon dioxide, although thankfully not nearly as abundant.

You wouldn't think the meagre human population millennia ago would be sufficient to cause global effects, but Ruddiman argues that time is the answer. Yes, it's true that in the late 1700s and early 1800s the amount of CO_2 released was nowhere near that reached when industry really kicked in by the late 1800s. However, in this case we're not talking about half a century, but fifty, sixty, even eighty centuries' worth of greenhouse gas emissions, although at very low rates.

But is there any substantial evidence that forest clearing reached such dramatic levels? There is. The amount of pollen from trees had already declined significantly in China by six thousand years ago. The Domesday report commissioned by William the Conqueror in 1086 revealed that less than 15 percent of England was still covered by

> We're not talking about half a century, but fifty, sixty, even eighty centuries' worth of greenhouse gas emissions.

natural forest. When you consider that clearing land for agriculture had been going on for centuries in Europe and Asia before it began in England, deforestation must have been widespread and profound.

Having concluded that archaeological evidence supports the idea that the clearance of forests for farming (starting eight thousand years ago) might account for the slow rise in the CO_2 record, Ruddiman then had to account for the opposite effect: odd, occasional, but significant reductions in carbon dioxide. Like the ice core that records a drop of about ten parts per million of CO_2 between 1300 and 1400. As far as Ruddiman is concerned, the only possible cause is the Black Death, which reached its peak in Europe between 1347 and 1352. An epidemic that kills half the population has predictable results: farms are abandoned, trees encroach on the land, and within a few decades, the maturing forest is a carbon dioxide sink, and the global carbon dioxide increase is halted, at least temporarily. And the effect *is* temporary; as the human population is re-established, the land is cleared again.

Ruddiman's thesis is controversial, and the ice core calendar unfortunately is not precise enough to prove this is what happened. But it is a provocative thought, one that, if nothing else, should help erode the feeling some of us have that we are too insignificant to affect the vastness of the earth's atmosphere.

▲ The cultivation of rice began about five thousand years ago. William Ruddiman thinks that was one of the crucial first steps in the generation of greenhouse gases and human alteration of the climate.

SIX

Driving Us Crazy

In 2008, the Progressive Automotive X Prize
competition was announced, a $10-million contest
to design the best and most practical car that
gets the equivalent of 100 miles per gallon of gas.
The X Prize should be an important step toward
designing vehicles that leave much lighter carbon
tire-prints. It's a serious contest, but even with
constraints, a carnival of vehicles will likely take part.
Think of what would happen if the constraints were
removed and car-lovers' ingenuity ran amok …

THE UNO

EIGHTEEN-YEAR-OLD BEN GULAK IS AN inventor. He wanted to build a small, environmentally friendly electric (no fossil fuel consumption) vehicle, but one that was attractive, even cool. And he came up with this: the Uno (pronounced "you know"), a cross between a street bike and a unicycle, but with a motor. Actually with two motors. It's a done deal now, but believe it or not, a little more than two years ago the Uno was no more than a school project. Now, it's a vehicle with a very sophisticated design.

▶ The Uno: steers like a Segway, goes a lot faster, and looks like nothing else on earth.

▲ It's a small, environmentally friendly vehicle, precision engineered, and guaranteed to turn heads.

"I wanted the bike to be as small and compact as possible," Gulak says. "And the best way of doing that is, instead of having wheels in tandem like a conventional bike, having the two wheels parallel, side by side. So you get the compactability of a unicycle. Now in order to do that, and have both motors fitting inside the frame, I had to offset the motor so that each wheel is driven by its own independent motor."

Gulak is packing a lot of equipment into a very small space: "It's so precise that in some places there's only a few thousandths of an inch clearance between what's moving and what's not moving. Almost touching. Precision engineered, that's what it is."

Everywhere you look on this ... this Uno, there are surprises. The electric motors and batteries aren't so out-of-the-ordinary, but the driving controls are. Because there aren't any! "Everything is

The Uno was no more than a school project. Now, it's a vehicle with a very sophisticated design.

▲ The power and brains of the Uno revealed.

done by leaning forward and backward and side-to-side," Gulak explains. "As the riders lean to one side, the wheels slide up and down inside the frame, so the whole bike can tilt to that side."

So, lean left, lean right, and the Uno will turn. But what if you lean forward, or back? Gulak has the answer for that too. "These gyroscopes detect the bike's forward lean and backward lean. As the rider leans forward, the motors speed up and the bike accelerates off. The same for going backward, and side to side. It's all the same idea."

Gulak's bike is somewhat reminiscent of the Segway, the two-wheeled vehicle that responds to body movements. But the Uno has something that the Segway doesn't: it is stylish. And most of all, it's smart: "You don't have to have a good sense of balance … you don't have to know how to ride a unicycle. The bike does all the thinking for you. You just sit on it, and lean."

Gulak has no idea whether the Uno will take off or not. At the 2008 Toronto Motorcycle Show, it drew amused, confused, and bemused glances, but most of the passersby liked it, if somewhat cautiously. And they appreciated that it was a cool, environmentally friendly idea.

That is a high-performance car …
But would any have a better stat than this:
95 percent recyclable and biodegradable?

BRITISH RACING GREEN

WHAT DO CASHEWS, CANOLA OIL, AND HEMP HAVE IN common? A not-so-memorable sixties get-together? Uh-uh. How about a green racing car? That may sound unbelievable, but it's true. And this is the car: the Eco One.

Designed by engineers at the University of Bristol in England, the Eco One is a rare combination of speed, power, and sustainability. Ben Wood is project manager: "We estimate the top speed to be 140-plus miles per hour, maybe 150. Zero to 60 miles per hour comes up in just around three seconds. That's a very, very fast accelerating car. It has a power-to-weight ratio better than a Ferrari Enzo."

That is a high-performance car, but let's face it: plenty of those have already been built, and many would likely have better stats. But would any have a better stat than this: 95 percent recyclable and biodegradable? I doubt it. And making it so was a challenge.

"The number one problem we encountered was trying to replace glass fibre with something a bit friendlier to the environment, something more natural. Glass fibre can't be recycled, and you can't even burn it because the glass leaves a horrible mess in the bottom of any incinerator you use. Nor does it biodegrade in a landfill—it stays around for a long, long time."

That's where the hemp comes in.

"We found some hemp fibre that's double-needled, like a felt basically: hemp fibres pushed together. And we developed a resin made from vegetable oil, canola oil, which we turned into a plastic, a two-part polyurethane resin, and we used those to make the body. When you mix them all together in a chaotic way, you have good strength in all directions."

◀ There are cars that get good gas mileage and cars that emit few pollutants, but very few like this one: it's 95 percent recyclable!

▲ Yes, that's hemp all right: it's in food, clothing, and now the body of Eco One.

The hemp matting together with the resin gradually solidifies into the body. But if you're a conscientious car owner, you know that the ability to accelerate, while admirable, isn't nearly as important as the ability to stop. And here the Eco One took another unexpected turn.

"A cashew nut comes in a great big husk the size of a baseball. [The husk] can be squeezed and squashed down to make a good resin which is very good at resisting temperature. We mixed that with a variety of other plant fibres and compounds, pressed it together in a mould, and baked it in an oven. It actually makes a high-performance brake pad."

It only gets better. How about tires? According to Wood, "They can be used on any car and they are about 100 percent potato starch. They're made by one of the major tire companies and are road legal."

And it doesn't stop there: "If you chopped up a potato and left it out in the air it goes black after a while. You can use that to tint the tires to make them a nice, dark colour. That replaces any fossil fuels or carbon black, products usually used to tint the colour of tires. The potato starch can be made very sticky as well. Because of that the tires have a good amount of grip but lower rolling resistance, meaning that the car uses less fuel, creates fewer CO_2 emissions, and less CO_2 is used in the manufacturing of the tires."

Lube the Eco One with a blend of plant oils—completely biodegradable oils, that is—and you actually get better lubrication and encounter less resistance in the engine. But if you spill *this* oil on the floor, a month later it has virtually disappeared. And the car runs on e85 fuel, 85 percent ethanol and 15 percent gasoline. It's too bad that ethanol has become so controversial because of its environmentally hostile production process—biofuels could and should be good replacements for fossil fuel, and likely one day they will be. Then the Eco One can maintain its nearly flawless credentials, at least until Eco One Squared comes online, a proposed 97 percent recyclable successor to Eco One.

"We're hoping to use a carbon fibre, a plant fibre found in wood to make a monocoque chassis like a Formula 1 car, but out of plant-based material. Biocomposites are what we call them. Plant materials, plant fibres, and bio-based resins out of carrots, potatoes, maize, sugar beet, sugar cane. That's one of the purposes: to show that these materials are safe to use and can be high performance as well."

It's not just that we should expect industry to make use of some of these "new" materials. It's that it would be great to see a Green F1.

▼ The tires are 100 percent road legal potato starch. They're even dyed with potatoes that have turned black.

▼ Is that a cashew or a brake pad? It's both!

Polar Bears

Of all wildlife that is likely to be or already has been affected by global warming, the polar bear is definitely the most photogenic and captivating, more iconic even than the penguin. And most people know the general outlines of the bears' story: they depend on sea ice, because it gives them access to the seals they prey on. As the sea ice in the Arctic melts or recedes from shore, bears are suffering from the lack of food, even starving to death. Bears are drowning because they now have to swim long distances. Bears are even, according to some reports, resorting to cannibalism. But as with any account of species at risk, you have to be very careful that there is a good ratio of data to speculation.

Polar bears are found all over the Arctic: 25,000 in all, 15,000 of those in Canada and Alaska, the rest scattered through Russia and Norway. In Canada the bears live in Nunavut (apparently numerous enough to support a trophy hunt every year, for which hunters pay tens of thousands of dollars), across the Arctic, and around the shores of Hudson Bay. Those Hudson Bay polar bears, especially the ones that hang around Churchill, Manitoba, are well known both to tourists and scientists.

But just east of there, on the shores of James Bay, is the world's southern-most population of polar bears. Martyn Obbard is a scientist with the Ontario Ministry of Natural Resources. He has collected data on these bears, and those data convince him that polar bears are seriously threatened by climate change.

▲ Waters like these, largely ice-free, are worrying.

You can't exaggerate the importance of actual data. The polar bear's state of health is controversial: sometimes local residents dismiss the opinions of scientists because they visit only occasionally, while they, the residents, are out there on the ice all the time. That's reasonable enough, but when you've got actual numbers you deserve to be listened to.

Obbard gets his data by following bears from a helicopter, darting them, then while they're anaesthetized, taking samples, measuring, and weighing them. Five years of doing this has revealed that something serious is going on: "There has been a significant decline in body condition. Animals, in general, are about 15 percent lighter for a given body length than they were in the mid-1980s."

This is not somebody's uneducated guess: the bears *are* skinnier, and that can have serious implications, as Obbard explains. "Ultimately, it will have—it *has* to have—effects on reproductive success and beyond that, on survival. If the summers continue to get long here, the bears are going to get skinnier until they reach a breaking point." Obbard thinks that warmer weather has shortened the bears' hunting season by about a month, enough to put them into nutritional stress. And he can say with some confidence that they will reach a

▲ Martyn Obbard of the Ontario Ministry of Natural Resources.

breaking point, even though right now his bears are maintaining their numbers. The bears near Churchill have been monitored for years by scientists from the Canadian Wildlife Service who have done the same things as Obbard is with his James Bay bears, but for much longer. They too saw an initial decline in weight without loss in numbers, but later the numbers started to fall. What reason is there to think the same won't happen in James Bay?

Although this point may be moot if the population shrinks dramatically, Obbard is now putting GPS collars on female bears to see exactly where they spend their time during the mating season, in April and May.

If they're unable to move easily across the ice at that time, the population might break up into smaller and smaller isolated groups. That is never good for a species' future: it needs to mix and match to maintain its genetic health. At the time of writing, a committee in Canada recommended classifying the bear as an animal of special concern and talked about the possibility of establishing a management plan. But the United States has recently classified the polar bear as a threatened species, with good data to support that decision. In northern Alaska, studies from 2001 to 2005 showed a clear correlation between the amount of ice and the success of the bears. In the first three years, the near-shore ice melted for about 100 days, and the polar bear population grew about 5 percent per year. But in 2004 and 2005, the number of "ice-free" days increased abruptly to about 135. Instead of increasing by 5 percent as usual, the bear population *declined* by about 25 percent per year.

But even if both the United States and Canada set out to "manage" the polar bears, what could they do? Besides not killing them, there's not much. If they're failing because of climate change, there's nothing we can do for polar bears except slow global warming.

"If the summers continue to get long here, the bears are going to get skinnier until they reach a breaking point." MARTYN OBBARD

▼ Data gathering like this has shown that the James Bay polar bears are in decline.

The Harlequin Frog: A Small Amphibious, Tropical Polar Bear

The case of the harlequin frog in Costa Rica is a perfect example of how difficult it is to determine just what effect global warming is having on any species of animal. These frogs (which are actually misnamed toads) are tropical; they live in mountain rain forests in Central and South America. They merit the label "critically endangered," as their numbers have been falling steadily over the last two decades and some species have disappeared from their natural habitat.

There has been controversy over what is killing these frogs, with global warming being one of the chief suspects. But in 2006 a group of researchers announced they'd put the pieces of the puzzle together. Frog die-offs seemed to happen during warm years, and they had found a fungus that looked like it was the cause of death, a fungus that couldn't have been thriving in the frogs' habitat unless it was becoming drier and warmer. The fungus probably lived there in low numbers before, but climate change had made it more virulent.

But not every researcher was satisfied this was the final answer, and in 2008 a different group presented evidence that they said showed the fungus had been introduced into South America in the 1980s and had spread through the mountains ever since. Sudden die-offs of frog populations would then be the result of the recent arrival of the fungus. The group found no evidence that these die-offs were linked to global warming. Their claim was that the fungus caused sudden declines of frogs in some areas because it had just arrived there, not because warmer temperatures had allowed it to flourish suddenly. This was a case that looked to be iron-clad evidence of the impact of global warming on a species—some even called it the first such evidence. Now, we're not so sure.

HUMMERS
IN REHAB

IT MUST BE THE MOST OBVIOUS TARGET for environmentalists, the supreme example of disregard for the earth. Although mystery shrouds the exact gas mileage achievable with a Hummer, everyone knows that it is, if not the most gas-hungry automobile in North America, certainly within the top two or three. Of course, Hummers' off-road capabilities are the reason people buy them, right? If off-road includes the driveway, then yes. But in a world where green ideas are taking hold, and the sense is growing that we just might be in some sort of trouble, even the Hummer is not beyond redemption.

And here is where that happens: Intergalactic Hydrogen, run by the father and son team of Fred and Tai Robinson. Fred has been tinkering with hydrogen-powered cars and boats since the 1970s. Tai was born into the business. Fred and Tai together can turn almost any vehicle, no matter how reprehensibly unenvironmental, into

"The vehicle you
drive is not as
important as the
fuel you choose."
TAI ROBINSON

something sweet. A prime example is their H2 Hummer, a car that can run on five different fuels, including hydrogen. "It's the same engine that came with the truck," according to Tai. "The internal combustion engine is at the top of its game—it's highly refined, it works very well, and it will run multiple fuel sources through it. All you need is the fuel storage and the fuel delivery system."

Fuel storage ... hydrogen ... isn't that what did in the Hindenburg? How come nobody worries about the safety of a tankful of extremely flammable hydrogen gas in the trunk of your car? On this subject, Tai is bullish: "The storage tanks for natural gas and hydrogen are bulletproof. They're rocket launcher proof. If my truck were run over by a train, the only thing left would be the tank. And the tank would still be holding pressure."

And, Fred says, there's more: "I also use that hydrogen to blend into natural gas to make what we call Intergalactic gas. Just a small portion of hydrogen in the natural gas reduces the emissions in the natural gas and increases its power significantly."

One of the neatest things Fred points out about this car is that you can switch from one kind of fuel to another while you're driving: "There's an interface box and it just switches from one system to the other. Right now we're running on natural gas. A flip of the switch: now it can run on ethanol. You

▲ This engine hasn't had a whiff of gasoline for thousands of kilometres.

didn't feel a thing; you didn't hear a thing."

Hydrogen, natural gas, ethanol, gasoline—the Hummer can use them all. And while hydrogen is what excited Fred back when he started, and it's the absolute cleanest with only water as exhaust, even the natural gas—especially the Intergalactic version—is a superior fuel: "When you burn natural gas in an internal combustion engine, the air coming out of the tailpipe is cleaner than the ambient air that people are breathing in downtown Los Angeles or New York or Houston."

Money is no object, or at least that's the attitude you need to have if you want your Hummer upgraded to burn hydrogen. The price could reach into six figures, but just think: you could then be just like California governor Arnold Schwarzenegger, who was given a hydrogen-powered Hummer by GM. Well, actually GM still owns and maintains it, but they let Arnold drive it. This effort by GM to make a Hummer green and then lend it to the guy with the high profile was, as Dan Lienert put it so nicely in *Forbes* magazine, like "putting lipstick on a pig."

Tai and Fred are different: they've combined their fascination with, and love of, the internal combustion engine with an honest desire to drive cleaner and more efficiently. So much so that they've entered the Progressive Automotive X Prize.

▲ Tai calls it a "steam bath." That's the beauty of hydrogen fuel: no emissions except good old H$_2$O.

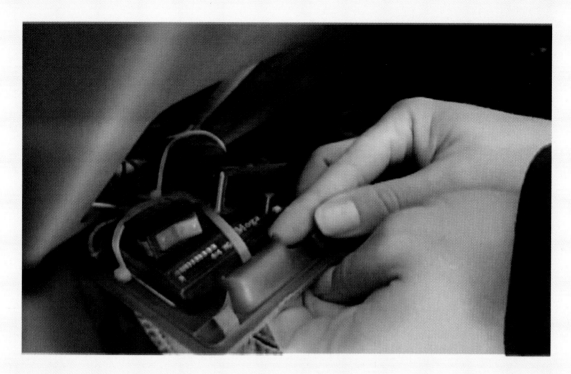

▲ The H2 Hummer runs on five different fuels with the flip of a switch.

WE WON'T GET THERE FAST, BUT WE WILL GET THERE CHEAP!

THERE ARE ALL KINDS OF CAR CLUBS: VINTAGE, CLASSIC Porsche, you name it. But the club that is in tune with the times, that can take $100-a-barrel oil and just laugh, is the Hypermilers club. The founder is Wayne Gerdes, a man who can squeeze 180 miles per gallon out of a car designed to get only 66. Gerdes is the champion Hypermiler.

Until ethanol, hydrogen, or plug-in hybrids rule the roads, the best thing any driver can do to limit carbon dioxide emissions is reduce the consumption of the fossil fuels that generate them. Small, compact cars and hybrids help, but what is usually overlooked is the role of the driver.

Driving to save gas is Wayne Gerdes's specialty, and as far as we know, he does it better than anyone else on earth. It starts even before he gets in the car. Everything that isn't needed is tossed. Gerdes has even been known to shed his sneakers if he's about to drive in a gas-saving competition. He wouldn't mind losing some weight, either—after all, every kilo costs gas. He also makes sure his tires are pumped up to the max. We doubt he would endorse inflating them to the 75 psi of the Cadillac Coupe deVille tires in Hunter S. Thompson's *Fear and Loathing in Las Vegas* (the two front ones were "tighter than snare drums"), but in Gerdes's way

◀ Champion Hypermiler Wayne Gerdes.

"I try to use the 'drive without brakes' technique."
WAYNE GERDES

of looking at things, the harder the tires, the less the rolling resistance—it's physics to save the earth!

Emptying the car of everything redundant to good mileage is one thing, but once Gerdes is in the car, you might not want to drive behind him, or even with him, because his driving style is, to put it mildly, unorthodox. For one thing, keeping up to the speed limit is the last thing he's interested in doing. He keeps his eye on the fuel gauge, not the speedometer. So you'll find Gerdes doing 80 kilometres an hour on the highway when everyone else is over 100. That might be a pretty obvious fuel-saving technique (remember how U.S. speed limits were dropped to 55 miles per hour during the oil crisis of the 1970s), but it is only the beginning. Off the highway, things get really interesting. For one thing, Gerdes doesn't just drive with a steady foot on the accelerator; instead, he employs "pulse and glide": "Pulse and glide is basically you're accelerating from one speed at a lower speed limit. We're going top pulse up to about 34, 35 miles per hour and then we basically coast down into a lower speed, 28, 29, then we repeat."

This requires the kind of patience that can come only from extreme dedication to good gas mileage. But if you're thinking that this is the most boring driving clinic you've ever witnessed, hang on. You haven't seen Gerdes deal with stop signs or red lights. Actually, he does his best *not* to deal with them.

"I try to use the 'drive without brakes' technique. Now I will take corners a little faster than most. But I'm not going to hurt myself and I'm not going to hurt anyone else. Driving without brakes puts you in the mindset that you actually don't have any brakes. You increase your buffers, meaning the distance between you and the guy in front of you."

This is a man who never drives with the crowd. On the highway, he's crawling; on highway exits, he's 10 or 15 kilometres faster than the posted speed. No extremes, no waste.

Gerdes is using other techniques almost too numerous to detail, like "ridge-riding"—driving

▲ Is he engaged in DWB here, or is it P&G? One thing's for sure: he's fully involved in FE.

on the white line of the shoulder, which keeps his tires out of the water that accumulates in the very slight grooves worn in the road by countless other drivers, drivers whom you have to assume are heedless to the global climatic needs they are ignoring.

Speaking of climate, Gerdes is not fond of air conditioning, for obvious reasons: it simply consumes too much gas. He will close the windows on a hot day, he will open the vent as wide as it will go, but he won't turn on the a/c.

So if you're going for a ride with Wayne Gerdes, here are some tips:

1. If it's 30-plus Celsius, ask him if it's OK if you walk.
2. If the drive is more than 10 kilometres, bring a book.
3. If he asks you to chip in for gas, go for it!

THE WAYNE GERDES GLOSSARY

ICE: internal combustion engine

FAS: forced auto stop. This involves putting the car in neutral, turning off the engine, and gliding to a stop. The lack of power affects both power brakes and power steering, a risk that explains why this isn't legal in every jurisdiction.

DWB: driving without brakes (figuratively, not literally)

P&G: pulse and glide

FE: fuel economy. Appropriately, this *is* the last word.

OFF THE BEATEN TRACK

If You Can't Find It in the Supermarket … All the Better!

Fergus the Forager

The 100-mile diet has become a fashionable way of coming to grips with the enormous consumption of energy involved in bringing out-of-season and out-of-country foods to your table. British Columbians Alisa Smith and James MacKinnon came up with the idea in 2005, and its ripples have been spreading ever since. The idea is simple: the more foods you eat that are produced locally, the less stress you're putting on the planet (by avoiding burning fossil fuels in ships, planes, or trucks to get the food from where it's grown to your plate).

It's an uphill battle to convince people that this is something they should do—even Alisa and James like olives and chocolate (although not necessarily at the same time), and you can't grow either near Vancouver. But by making careful choices, you can reduce the carbon footprint of your diet dramatically. Even so, it's still common to read comments by chefs like, "I use Alberta beef most of the time, but if I can get Kobe from Japan, I'll go for it." So, somewhat disappointed by examples like that, we at *Daily Planet* decided that we needed to present a more hardline example than the 100-mile diet. In fact, it is pretty meek compared to Fergus the Forager.

Fergus Drennan is his name, and he puts all of us to shame by gathering almost every-thing he eats directly from nature. Forget the 100-mile diet … he doesn't even shop for his food. He *collects* it.

As he says, "eating seasonally and locally, the ultimate step there, even beyond organic, is wild food—you're just taking as

▶ Fergus "the Forager" Drennan sampling his collected wares.

244

"Sometimes it's hard to remember where your dinner is …"

FERGUS DRENNAN

much as you need on a daily basis, there's no packaging, and you don't have any of that guilt about being part of the horrible cycle of destruction."

Fergus lives in England, and even in that country, where truly wild and undisturbed lands are relatively rare, he has no trouble finding a wide variety of edible and even delicious foods that he can corral for free. It's estimated that tens of millions of birds and tens of thousands of foxes and badgers are killed on English roads every year. Even though Fergus estimates that roadkill makes up 5 percent of his diet, at most, he estimates that more than two million meals could be created from those carcasses. After

all, roadkill hasn't been factory farmed or pumped full of antibiotics. It is fresh, local, seasonal, and nutritionally rich—as long, of course, as you gather it within a day, or preferably a few hours, after it has died. "I've eaten badger, fox, moorhen, duck, seagull, hedgehog, rabbit, hare, owl, wren (believe it or not there was about a teaspoon of breast meat but I thought I'd try it anyway) … and, oh yeah, squirrel and pheasant."

However, plants were on the menu when Fergus set out into the countryside with a *Daily Planet* camera in tow, ready to demonstrate just what could be gathered in a day to arrive on the dinner table at night. He noted when picking the wild garlic that

▲ Never has so much free food been within reach.

it comes up much earlier—weeks earlier—than it used to, likely because the temperatures are going up. The sap is running earlier too in the birch trees he taps to make syrup. These are things that he has to notice, because many of the foods he gathers are only really delectable for a relatively short time in the spring.

It's not just timing that's important. It's awareness. As Fergus walks past what most of us would dismiss as a wasteland (that's exactly what it is called), he points out that it is full of edible greens, including rosebay willow, a herb that gets bitter later in the year but in the spring is, at least in Fergus's mind, "lovely." And then there's hairy bitter cress: "It's not bitter and it's not particularly hairy, so why they call it that I don't know, but it's really delicious and I think I'm just going to have to eat the whole thing."

Although much of this trip has admittedly been eat and run, the foraging lifestyle can verge on the elegant. Fergus left the wasteland to head to the beach, where he harvested some seaweed to wrap around the pike he caught earlier, so that he could bake the fish in a newly dug sand oven. He'll return to savour the fish in about twenty minutes, but to pass the time, he gathers—and snacks on—wild spinach. "You'll never want to return to the normal spinach."

There are risks: you don't want to fool with roadkill if there's a chance that it has been dead for too long, and while many mushrooms are absolutely delicious, as the saying goes, some of them are delicious just once—because they kill you. Only the very knowledgeable should collect and eat mushrooms.

AN EXPERIMENTAL
SOLAR CAR

ON A GOOD DAY OF CONSTANT TRAVELLING, THE LAST THING this car needs is a gas station (unless it's for the bathroom), or an electric plug, or in fact anything but some sunshine. This is the University of Calgary's solar car, Schulich 1 (named for the Schulich School of Engineering). It came first among other Canadian solar cars in the Panasonic World Solar Challenge in Australia, and placed twelfth overall out of forty-six. Its stats were impressive: 2999 kilometres travelled (from Darwin to Adelaide) in fifty-one hours and forty-three

◀ Sleek, refined, and totally dependent on the sun, the University of Calgary's solar car, Schulich 1.

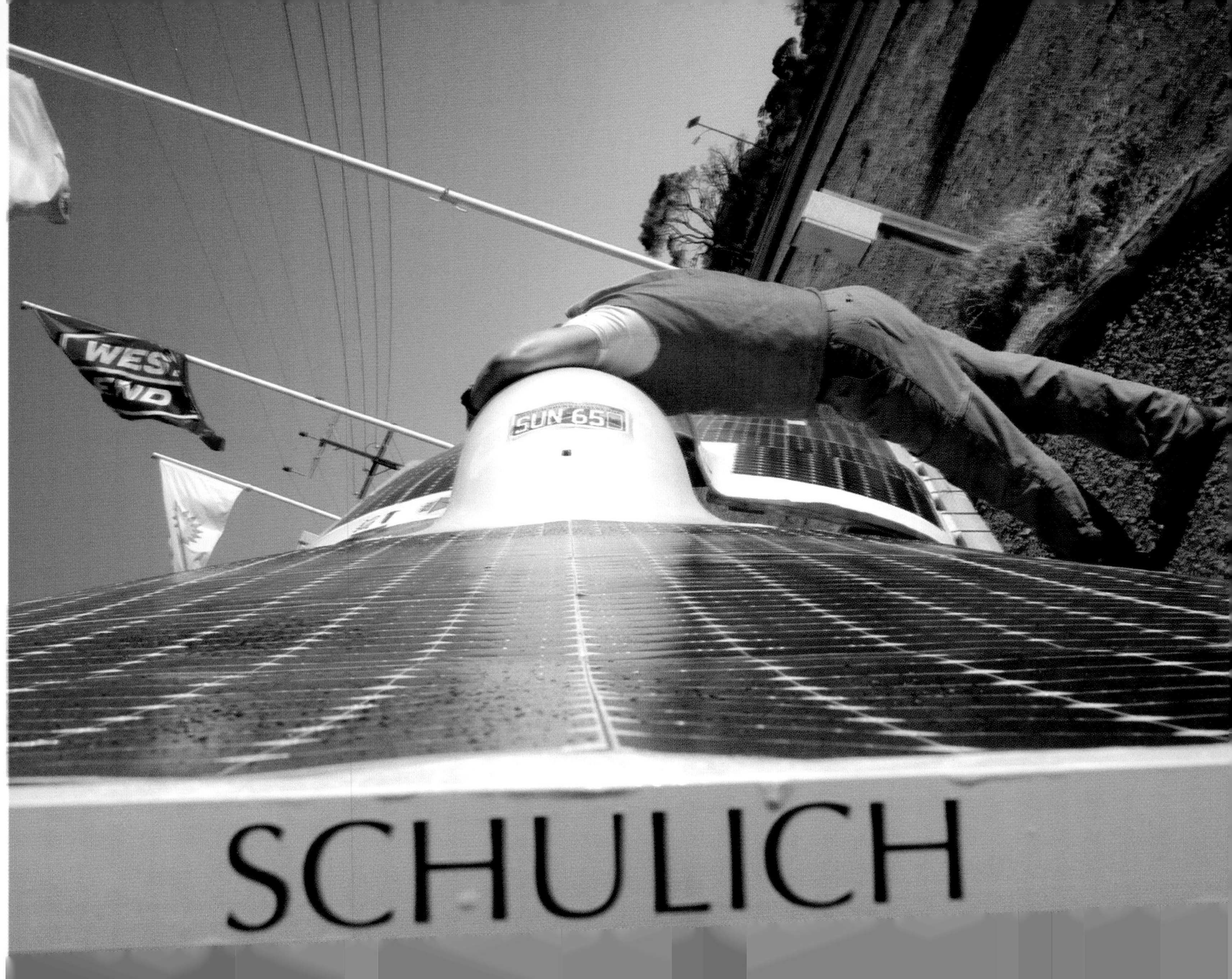
SCHULICH
SUN 65
WES
END

minutes, an average just short of 60 kilometres an hour. And the car had been damaged in a minor accident that no doubt slowed it down.

Solar cars like this one are a bit of a contradiction. On the one hand, they surely represent the ultimate advance in reducing the environmental impact of cars on the planet. Solar is free, after all. But the solar cars that participate in races like the Panasonic in Australia or the North American Solar Challenge are not exactly street legal. They are taken to extremes to be able to operate on the sun's energy only, and those extremes make it unlikely that anything quite like these could become consumer products. But who cares? They are a beautiful example of ingenuity and innovation.

The basics: Schulich 1 has a single driver and three wheels. Although it's capable of 120 kilometres per hour, 80 is a more reasonable average. And it's huge: 6 metres (18 feet) long. As you'll see, the car is designed not to be comfortable, but to be efficient in every conceivable way. It has to squeeze every bit of useful energy out of its solar cells, and put as much of that energy to work turning the wheels as possible. That is what it's all about. But as simple as that is to state, it's devilishly difficult to achieve.

It's tough for drivers even before they get into the car: gym three times a week, classes on race driving techniques, even "hot yoga" classes to prepare them for the heat.

◀ Every single source of friction, resistance, or inefficiency must be scrubbed out of the car to enable it to win races.

So, for instance, the shape must be aerodynamic. But because this is a student project with limited resources, the designers can't afford to rent the time in a wind tunnel big enough to accommodate their car. But they know their car isn't very aerodynamic, and there isn't time to build a completely new one. They're stuck with what they have. So they have to resort to the next best thing to a wind tunnel, a combination of measuring how the car does at different speeds, and at the same time referencing a software package called Fluent. This software can figure out the flow of water over the body of an elite swimmer, the currents of air around a cornering motorcycle, or, in this case, the slipperiness of a solar car. Together, these measurements will allow the team to decide exactly how to get the best out of Schulich 1.

▼ The Schulich 1 is designed to be aerodynamic; the driver's comfort comes second.

Aerodynamics are one thing; material, another. The lighter the car the better, so the shell of Schulich 1 is made of Kevlar and carbon fibre, the chassis of chromoly, a very light steel. The solar cells are triple-junction gallium arsenide cells, relatives of the cells used to power satellites. The captured solar energy is channelled through the array of solar cells and used to charge a battery pack, 30 kilograms' worth of lithium polymer batteries (higher quality than the ones they ran on in Australia).

All these factors are important, but so is the driver. Racing in Australia showed the Calgary team just how difficult it is to keep the driver comfortable. Darshni Pillay, co-chair of the project, explains, "The first thing that we learnt was that the car wasn't very driver-friendly. It is not ergonomic at all. We had an issue with the driver's legs falling asleep after a few hours. We came up with a short-term solution—a couple of

▼ Kevlar, carbon fibre, chromoly triple-junction gallium arsenide solar cells: this the stuff of a winning solar car.

cushions—but this is something that we are working on improving on. Also, it doesn't have any back support.

"The fact that it's a three-wheeler makes it handle very differently from the usual four-wheeled vehicle, and it's not power steering, so your arms get tired. Then it's even more challenging to stay in control of an already tricky vehicle. And it gets hot: solar car races tend to be held in places where the sun is beating down on you. All of this adds up to a tough five-hour shift for a driver."

It's tough for drivers even before they get into the car: gym three times a week, classes on race driving techniques, even "hot yoga" classes to prepare them for the heat. At the time of writing, the team was aiming for the North American Solar Challenge, starting in Texas and finishing in Calgary. As that race approaches, the drivers will increase their water intake to prevent dehydration.

In the end, it won't so much be the cars of the solar races that take us into the sustainable future as the students who design them.

The Moral Hazard of Engineering Climate Change

David Keith

David Keith is at the Schulich School of Engineering at the University of Calgary. He, like Klaus Lackner (see page 88), is developing technologies to scrub carbon dioxide from the air. This 6-metre-high tower is a prototype of the kind of carbon removal unit he would like to be able to deploy. This unit can take the carbon dioxide from ten cars out of the air. It's not much; it seems like an elaborate piece of equipment for just ten cars. But in Keith's mind, developing technology to achieve carbon-neutral transportation is so important we must explore some long shots.

"If someone appointed me czar of energy tomorrow, I'd know right away what to do about the carbon emitted in the generation of electricity. That is a pretty straightforward problem. I'd invest heavily in wind power, nuclear power, and carbon capture [at the smokestack] technologies. Solar is too expensive right now, but I'd invest heavily in the research and development of solar. But when it came to transportation, it's not so clear what to do. I don't think hydrogen cars make much sense ... I'm not convinced that hydrogen is going to work. They're technically possible but hugely far away. I'm more excited now

For scientists and engineers like David Keith, pursuing a single research project simply isn't enough.

than I was ten years ago about electric cars, but they still have a long way to go and there are real fundamental chemical limits to batteries. And of course we all know what the problems with biofuels are. That's what convinces me that this technique—capturing carbon dioxide out of the air and then using it to synthesize fuel—is worth a hard look."

For scientists and engineers like Keith, pursuing a single research project simply isn't enough. He has taken a leading role in organizing meetings around the world to put geo-engineering on the table, trying to make sure the controversial ideas—like injecting sulphate particles into the upper atmosphere—get a fair hearing, one way or another. Keith is fully aware that the mere possibility of geo-engineering can make the threat of climate impacts look less fearsome, which in turn would lessen our commitment to cut carbon emissions. This is called a "moral hazard." Despite this difficulty, Keith has been surprised at the rapidity with which opinion has moved toward at least seriously researching these proposed technologies. He has come to that point himself, although, as he says, "it's been a reluctant journey."

And is he optimistic that we will be able to do enough to prevent widespread fallout from carbon dioxide emissions and global warming?

"It depends on which side of the bed I get up on. I am very optimistic that technically this is doable. And that's not true of all things. You can't just pay to fix the spread of nuclear weapons. You can't just pay to fix complex things, like creeping deforestation. Because CO_2 is a simple gas, you can fix it. But on the other hand there's the fact that people knew about this problem for forty years, and did nothing, and then we all agreed we were going to do something about it, and yet the emissions are *accelerating*. Part of the solution is technological change, but there's also social change, and on that side it seems like we are getting nowhere. So I go back and forth on this."

A PLUG-IN ELECTRIC ON A TEENAGER'S PAY

ANDREW ANGELLOTTI, LIKE MANY TEENAGERS, WORKS as a lifeguard at the local community centre close to his home. Well, it's not exactly walking distance: it's about a ten-minute drive, just short of 20 kilometres round-trip. So he drives. Shaking your head sadly? Dismayed that here's yet another example of our dependence on cars? Well, you can stop worrying and start marvelling right now, because in Angellotti's complete round-trip he doesn't use one millilitre of gasoline. His truck is a plug-in electric. That's pretty impressive. But what takes it from impressive to amazing is that he did the conversion himself.

"The truck's capable of going about 55 miles per hour on a good day," Angellotti explains. "I usually drive it around 45 or 50 just to be easy on the batteries. I bought the truck intending to convert it to electric power. I bought it when I was fifteen and spent about nine months converting it. I love it."

He was sixteen years old when he converted a standard internal combustion engine to plug-in electric? And it only cost about $6000? These days the idea of plug-in electrics seems to be just around the corner, or maybe not even "just," but around that future corner somewhere. Yet here's a kid who did it himself. How did he afford all that ultra-high technology on a part-time

▶ He's just a neighbourhood kid, but he's outdone the big carmakers.

258

259

lifeguard's salary? Well, it turns out the technology wasn't really like that.

"This is hair dryer technology," Angellotti explains. "I mean the technology I used in my truck is exactly the same as the components that someone would have used to convert twenty years ago. That's what I could afford, but there are technologies out there today that could do a lot better conversion."

It's true—even the dashboard of the car isn't that radical. There's an amp meter that gives you an idea of how fast you're using the electricity stored in the batteries, and a voltmeter that, like a fuel gauge, tells you how much power you have left.

Angellotti just plugs the car into the house electrical system to charge the batteries. Not only is it convenient, easy, and cheaper than buying gas, it's also his parents' house and it's their electricity bill.

In the end, Angellotti just thinks electricity is the way to go: "Hybrids still use gasoline; they get better gas mileage but they still use quite a bit of gas. They're a good option, but I like mine better. Right now I'm working on a 1992 Toyota Tercel. It's a lighter car, has a larger motor, higher voltage. So even though the batteries will be lower capacity to save space and weight, it should get really good performance."

It does make you wonder why we don't see more plug-in electrics, doesn't it?

◄ In the end, the much-vaunted plug-in hybrid was a spare-time project—for a sixteen-year-old.

▲ Andrew Angellotti at his day job.

"The main reason I wanted to do this is just to show people it can be done."
ANDREW ANGELLOTTI

THE PRODUCTION SOLAR CAR

IT'S CALLED THE VENTURI ASTROLAB, IT'S going to cost something like $100,000, it resembles a Formula 1 race car, and it is actually kind of amazing. Unlike Schulich 1, the solar car built by engineering students at the University of Calgary, this solar car is intended for use on the streets. Yet the resemblance between the two speaks to their evolutionary relationship. As president of Venturi, Gildo Pallanca-Pastor says, "The solar cars in those races are really the pioneers. We wanted to create the real car, taking the best of those incredible prototypes but making something that would be street legal. In the end, it's true that the Astrolab is a direct descendant of those cars."

The Astrolab isn't for soccer moms or even for anyone who goes grocery shopping. It seats two, snugly, one behind the other in a race car–like cockpit. The spare racing chassis is obscured by a giant sheet of solar panels, which provides a unique driving experience: "When you are in the car you feel like you're in the middle of a mirror, like a flying carpet. It's also silent, so driving is a really interesting sensation."

▶ You can see the Venturi Astrolab owes much to student solar cars, but somehow it's different.

▲ From the driver's seat, there is indeed the feeling that you're driving a flying carpet.

Like all solar cars, even the student-built models that compete, there are trade-offs everywhere. It has to be very light, so the structure under the solar panels is spare. But at the same time it has to be safe, so the Astrolab was designed with a Formula 1 type of crash box in the front and a very stiff structure all around. In the end, the Astrolab is light enough that it can get up to 120 kilometres per hour, and you don't have to worry about cloudy days—an onboard battery will store electricity.

As Pallanca-Pastor says, this is a car of virtue, completely sustainable as it is, but also, "It's not only good to be environmental, it has to be fun. And that's what we wanted to show."

Want one? Pallanca-Pastor isn't so sure this is the car for the Canadian climate. "Even though Canadian people are very tough, I don't think it's going to be the best car to go around in wintertime." But nonetheless you might be able to buy one in a year or two. Estimated price: about US$100,000.

▲ The Astrolab is more street ready than the solar cars engineering students make, but it's very expensive as a result.

"It's not only good to be environmental, it has to be fun."

GILDO PALLANCA-PASTOR

Acknowledgments

This book might have set some sort of record for the number of people who made significant contributions to it, but one person stands out. Ashley Bristowe was, as she puts it, "managing research editor" of this book, but actually she *ran* it. If there's anyone out there who's better at cajoling, researching, challenging, organizing, record-keeping, or working 24/7 to keep a project on track, I'd like to know who it is. (She's pretty funny too.) Without her this book would simply not exist.

Then there are the producers of *Daily Planet,* the people who sought out most of the stories in this book and created excellent television from them. I acknowledge their hard work and creativity, as I'm sure they would the efforts of others who helped them realize those stories. So, in no particular order, thanks go out to Cindy Bahadur, Stephen Sheahan, Shannon Bentley, Laura Boast, Agatha Kowalski, Seonaid Eggett, Amanda Buckiewicz, Doug Crosbie, Jeff Berman, Rob Davidson, Alix MacDonald, Penny Park, Kelly McKeown, Lori Belanger, Jeff Blundell, Mark Miller, Henry Kowalski, Steven Hunt, John Morrison, Michelle Chung, Carol McGrath, Frances MacKinnon, Jenn Sunnerton, Mark Stevenson, Nicole Sen, Ben Schaub, and Barb Ustina. A visual book needs visual guys, and *Daily Planet* has both Kevin Francisco and Mathew Knegt. A visual book also needs cameramen, and while there were innumerable ones involved in these stories, *Daily Planet*'s regulars, Jay Kemp and Ross Macintosh, were front and centre. There are, of course, many others who played important roles in getting the original television pieces to air (notably the editors), but I'm trying not to turn these acknowledgments into a chapter.

As in any book, there are innumerable people on the publishing side who were instrumental in getting the book out. Andrea Magyar at Penguin Group (Canada) was the person who first saw that a week-long television series on global warming could be the inspiration for a book, and Helen Smith, Tracy Bordian, and Marcia Gallego all put in serious time to make it a reality. On the Discovery Channel side, Anne-Marie Varner, Steven James May, Jennifer Preston, and especially Deborah James took care of the peculiar details that arise only when television is being converted to print and still pictures. Both Deborah and Paul Lewis played a crucial role by seeing the potential of this book and making it happen.

And finally, the scientists, inventors, and characters who populate this book made their most important contribution by agreeing to be part of it; some went further and offered assistance in the form of advice and even images. Among them are Gretchen Hofmann, Bea Csatho, Dan McCarthy, Julian Gutt, Larry Rome, Roberta Bondar, David Keith, and Bill Lishman.

Jay Ingram

Index

Credits

Grateful acknowledgment is made for permission to reproduce the following:

PHOTOS

Contents, page vi: Roberta L. Bondar, OC, O. Ont., MD, PHD, FRCP, FRSC

Introduction

Page viii: Jeffsrey Kargel, USGS/NASA JPL/AGU

Page xi (top): British Crown © 2008, The Met Office

Page xi (bottom): Professor Henri D. Grissino-Mayer, University of Tennessee

Pages xii–xiii: HIP/Art Resource, NY

Page xiv: Brian Bencivengo NSF/USGS National Ice Core Laboratory

Page xv: Figure 2-3, Contribution of Working Groups I, II, and III to the Third Assessment Report of the Intergovernmental Panel on Climate Change (IPCC), *Climate Change 2001: Synthesis Report,* published by Cambridge University Press

Page xvii: NASA

Page xix: SIO, U. California and NOAA/ESRL

Page xx: SOHO/MDI (ESA & NASA)

Page xxi: © Richard Harwood, Black Hawk College

Page xxii (top): Robert Simmon, NASA GSFC

Page xxii (bottom): Gary Hincks/Science Photo Library

Page xxiii: R.S. Culbreth, USGS

Page xxiv: Professor Dwight M. Smith, University of Denver

Page xxv: Jeffrey Kargel, USGS/NASA JPL/AGU

One: Extreme Science

Pages xxvi–1: Roberta L. Bondar, OC, O. Ont, MD, PhD, FRCP, FRSC

Page 3: *Daily Planet*

Page 4: *Daily Planet*

Page 5: *Daily Planet*

Page 6: Dr. Araki and Dr. Kahawara, Earth Simulator Center/JAMSTEC

Page 7: *Daily Planet*

Page 9: Julian Gutt

Page 10 (top): Julian Gutt

Page 10 (bottom): NASA-GSFC

Page 11 (top): Julian Gutt

Page 11 (bottom): *Daily Planet*

Pages 12–13: MacDonald, J.E.H. (Canadian, 1873–1932), *Early Morning, Rocky Mountains,* 1926. Oil on canvas, 76.2 × 89.2 cm. Art Gallery of Ontario, Toronto. Gift of Mrs. Jules Loeb, Toronto, 1977; donated by the Ontario Heritage Foundation, 1988. © 2008 Art Gallery of Ontario

Page 14 (top): *Daily Planet*

Page 14 (bottom): Data courtesy of the World Glacier Monitoring Service (www.wgms.ch)

Page 15: Data courtesy of the World Glacier Monitoring Service (www.wgms.ch)

Page 16: *Daily Planet*

Page 17: Daniel P. McCarthy, Brock University

Pages 18–21: Cryptic Moth Productions Inc.; photo by Gad Reichman

Pages 22–23: Brian Bencivengo NSF/USGS National Ice Core Laboratory

Page 24: *Daily Planet*

Page 25 (top): *Daily Planet*

Page 25 (middle): *Daily Planet*

Page 25 (bottom): Brian Bencivengo NSF/USGS National Ice Core Laboratory

Page 26 (left): Brian Bencivengo NSF/USGS National Ice Core Laboratory

Page 26 (right): *Daily Planet*

Page 27–31: *Daily Planet*

Pages 33–35: Courtesy of Gretchen E. Hofmann; photo by Nann A. Fangue

Page 116: *Daily Planet*

Page 117: Dan Vicroy

Page 118: *Daily Planet*

Page 119: Dan Vicroy

Page 120: *Daily Planet*

Page 123: *Daily Planet*

Page 124–127: *Daily Planet*

Page 128–131: *Daily Planet*

Four: Getting Around

Pages 132–133: Greg Ehlers, SFU Photographer

Page 135: Dr. Evan Goldman

Page 136: Greg Ehlers, SFU Photographer

Page 138: Greg Ehlers, SFU Photographer

Pages 140–143: *Daily Planet*

Pages 144–147: *Daily Planet*

Pages 148 (top): Courtesy Brian Peters and Smog Veil Records

Page 148 (bottom): Courtesy Doug Fogelson and Smog Veil Records

Page 149: Courtesy Saverio Truglia and Smog Veil Records

Page 150: Courtesy Greg Gibson and Smog Veil Records

Page 151: Courtesy Saverio Truglia and Smog Veil Records

Page 152: Courtesy Greg Gibson and Smog Veil Records

Pages 154–157: *Daily Planet*

Page 158–159: *Daily Planet*

Page 160: Bernie Krause

Pages 161–163: *Daily Planet*

Five: Where On Earth Will We Live?

Pages 164–165: Courtesy of Bill Lishman, photo by Peter Begg

Page 166: Peter Amerongen, President: Habitat Studio & Workshop Ltd.

Page 168: Ashley Bristowe

Page 169–171: Peter Amerongen, President: Habitat Studio & Workshop Ltd.

Pages 172–175: *Daily Planet*

Pages 177–178: Jay Shafer, Designer: Tumbleweed Tiny House Co.

Page 179 (top): Jay Shafer, Designer: Tumbleweed Tiny House Co.

Page 179 (bottom): John Friedman

Page 180: Ashley Bristowe

Page 183: Ashley Bristowe

Page 184: Ashley Bristowe

Page 185 (top): Jamey Stillings, President: Jamey Stillings Photography, Inc.

Page 185 (bottom): Ashley Bristowe

Page 186: Ashley Bristowe

Page 187: Cryptic Moth Productions Inc.; photo by Ian Connacher

Page 188: Cryptic Moth Productions Inc.; photo by Ian Connacher

Page 190: Courtesy of Bill Lishman, photo by Peter Begg

Page 192: Photo by William Lishman

Page 193: Courtesy of Bill Lishman, photo by Peter Begg

Page 194: Courtesy of Bill Lishman, photo by Peter Begg

Page 195: Photo by William Lishman

Pages 196–197: Courtesy of Bill Lishman, photo by Peter Begg

Page 198: Courtesy of Amory Lovins

Page 199: *Daily Planet*

Page 201: *Daily Planet*

Pages 202–206: Luke Everingham, Designer/Owner/Builder: Everingham Rotating House

Page 207: Rolf Disch, Architect

Pages 208–209: Luke Everingham, Designer/Owner/Builder: Everingham Rotating House

Page 211: Courtesy of Museo Nacional del Prado

Page 213: AP Photo/Apichart Weerawong

Six: Driving Us Crazy

Pages 214–215: University of Calgary Solar Team

Page 216: *Daily Planet*

Page 217: Glenn Roberts, *Motorcycle Mojo* magazine

Pages 218–219: *Daily Planet*

Page 220: *Daily Planet*

Page 221: Glenn Roberts, *Motorcycle Mojo* magazine

Pages 222–225: *Daily Planet*

Pages 226–227: FloridaStock/Shutterstock

Page 228: Jan Martin Will/Shutterstock

Page 229: *Daily Planet*

Page 230–231: *Daily Planet*

Page 233: Paul Zahl/National Geographic Image Collection

Pages 235–239: *Daily Planet*

Pages 240–243: *Daily Planet*

Page 245: Fergus Drennan

Page 246: Fergus Drennan

Page 247: *Daily Planet*

Page 248–255: University of Calgary Solar Team

Page 256: Ken Bendiktsen

Pages 259–261: *Daily Planet*

Pages 263–265: *Daily Planet*

Pages 266–267: Photodisk

Pages 276–277: Shutterstock

Pages 284–285: John Pitcher/iStockphoto

TEXT

pages 82–83: Excerpted from Roberta Bondar, Introduction to *Passionate Vision: Discovering Canada's National Parks*, Vancouver: Douglas & McIntyre, 2000, p. 22.

pages 113 and 169: Excerpted from George Monbiot, *Heat: How to Stop the Planet from Burning*, Toronto: Doubleday, 2006, pp. xx–xxi, 66.

page 198: "I don't do problems … I do solutions," Amory Lovins, quoted by Elizabeth Kolbert in "Mr. Green," *The New Yorker*, January 22, 2007.

This book has been printed on
100-pound Arbor Web Plus Dull 2 paper,
FSC-certified, 30% post-consumer waste,
from the New Page Mill.